SCIENCE and the CHURCH MILITANT

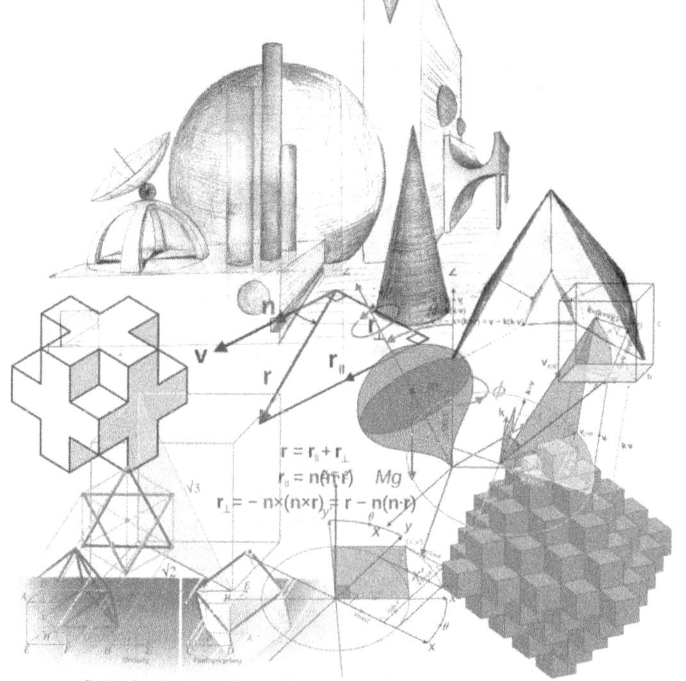

DYLAN J. THOMPSON, Ph.D.

Science and the Church Militant

Dylan J. Thompson, Ph.D.

Science and the Church Militant

By Dylan J. Thompson

ISBN: 978-1-64594-011-1

Published By Athanatos Publishing Group.

Table of Contents

About Me and About this Book

Essay 1	Truth, Beauty, and Inorganic Chemistry	1
Essay 2	Oda Nobunaga's Acceleration Due to Gravity	6
Essay 3	The Probability Density of Truth	13
Essay 4	Battle Lines of the Church Militant	17
Interlude		23
Essay 5	Wise as Serpents	25
Essay 6	The Self-Flipping Light Switch	30
Essay 7	Choose Your Own Authority	36
Essay 8	How Big was the Ark?	42
Essay 9	The Age of the Earth	46
Interlude		51
Essay 10	Science 'News' and the Philosopher's Stone	52
Essay 11	The Great Extrapolation	58
Final Interlude		65
Essay 12	The Enemy and His Heavy Cavalry	67
Bibliography		78

About Me and About this Book

During my childhood, my father took seriously Martin Luther's statement in the Small Catechism: '*As the head of the family should teach them in a simple way to his household'*, so he ensured that I was well-catechized. I was homeschooled my entire childhood and up through high school as well. During my 'junior' year of 'high-school' I heard an apologetics focused presentation by Craig Parton, and heard of the International Academy of Apologetics, Evangelism, and Human Rights in Strasburg, France. I attended the Academy after I 'graduated' from home-school at 17, in the summer of 2005, which was also the summer right before I went to college. I attended Hillsdale College, in Hillsdale MI, graduating with a B.S. in Chemistry followed by Purdue University in West Lafayette IN and graduated with my Ph. D. in August of 2014. I began teaching at Concordia University Wisconsin that same August. Since then, I have completed the teacher colloquy process with the LC-MS in April of 2018.

The following essays are all of a theme, yet they do not necessarily stack neatly upon each other. I believe that as you, dear reader, read (in any order you like) the whole idea will become apparent. It is like a painting by a pointillist, the idea is there, but it is for you to see it. One thing that post-modernists have right (despite the negativity I direct at that philosophy in the essays) is that individual people have to see the truth for themselves.

At the end of each essay there is a Study Work section. The goal of these sections is to give the reader, the group, or the class a direction to focus their further thought and study.

Essay 1: Truth, Beauty, and Inorganic Chemistry

This introductory essay is a broad outline of thoughts to share with you, my reader. It should convey a general feeling. I will spend little time defending my general ideas. That is for later essays: The goal is to provide a perspective into the worldview that will follow, and a blueprint for what to expect as you read. The references and counterarguments can wait. I hope to **spark a gut level,** (grok[1], so to speak) idea that the sciences and our Lutheran faith are not *fundamentally* at odds. In fact, they are in essence allies: science the lesser secular ally of the theologian who adheres to the Augsburg Confession. This collection of essays forms paving stones along the path toward acknowledging and engaging that alliance.

Well, introducing an introductory essay feels quite... meta. But alas, it could not be helped. Without further ado, I present the main attraction: Truth and Beauty alongside my favorite chemistry: Inorganic Chemistry. (Any old fool can see the hand of God in biochemistry... it takes a real philosophical grok to see it in Inorganic ☺) I have this simple definition of truth that I use (to be discussed at greater length in the essay handily entitled "Truth"): 'Truth is that one thing from which any deviation is falsehood". Truth is the needle-point of what actually is, and any misinterpretation at all is some level of falsehood. In all reality, though this definition is accurate, I have found it almost entirely useless in actual discussion. But one part of it, the most essential part of my working definition is this: Truth is one thing, and one thing only. There are no 'truths for chemists' that are actually different than 'truths for Christians', we may work with different

[1] From Heinlein, Robert, A. 'Stranger in a Strange Land' **Grok**: a Martian word which translates roughly 'To understand something to the depths of your being: to grasp an idea or event with your whole self, even your sub-conscious and soul.' (my summarized definition...)

approximations of the truth between an organic chemist who loves the valance bond theory of bonding and the inorganic chemist who insists on molecular orbital theory, but the truth itself is what it is, and it is one thing. There are fragments of the truth that we find that seem unrelated to each other, but ultimately, I believe that all truth shares the same characteristics because it is the same thing.

So when I now split truth into the two classic categories, revealed and discovered, I will continue to insist that the Truth itself, Truth qua truth, is one thing, everything that is true, both revealed and discovered, and that it cannot and will not contradict itself. All contradictions are due to the fallible human misinterpretations of something too enormous to understand. A theologian cannot legitimately claim to understand how the bread and the wine are actually the Body and Blood, and any attempt to explain will invariably fall to disaster. Likewise, the chemist cannot actually explain how it is that a 1s electron in a hydrogen atom could, technically, be *literally* anywhere except the nucleus, even across the room. But it is not… but it could be. Again, no good explanation. In the latter case, attempts at explanation do not risk heresy, so it is simpler, but the fundamental fact is that there are components of both revealed and discovered truth that escape the human mind.

Of course, I have a working definition of beauty as well. 'Beauty is an objective trait which elicits a subjective response upon providing a glimpse of unfallen creation.' It's a good one, right? What does that have to do with science, even Inorganic Chemistry? Well, let me show you something in Figure 1.

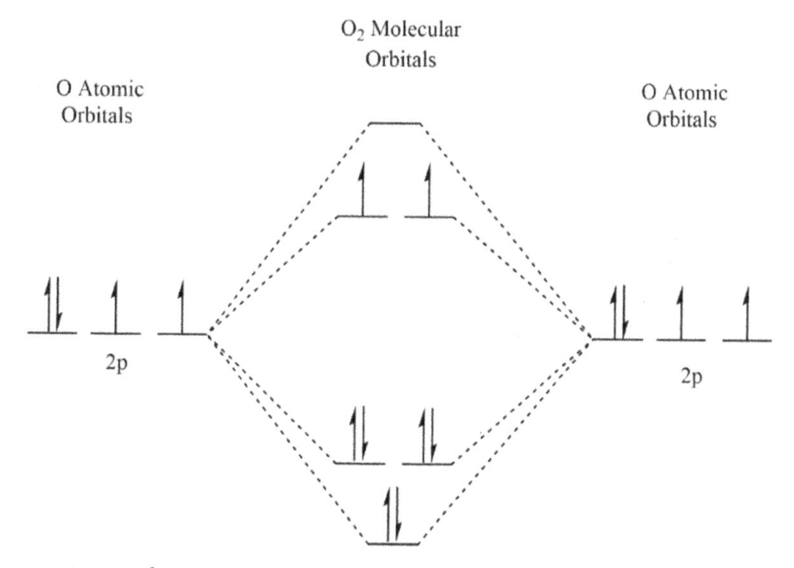

Figure 1:[2] Abbreviated Molecular Orbital diagram for O_2 bonding.

That was the molecular orbital diagram which accurately describes the bonding in the O_2 molecule. The other bonding theories are inadequate, and do not predict that O_2 is paramagnetic (attracted to a magnet). Seeing for yourself a representation of a glimpse into something that works essentially perfectly, despite the fall of creation, is an incredible *pathos* moment. Here you have seen and, if you take an inorganic chemistry class begin to grasp, how the bond is actually formed. It is as beautiful and as tragic as Waterhouse's Ophelia. It should not surprise that it requires a level of understanding to feel such a feeling, but consider how well so much of creation works, and how intricate it is and then the crushing burden of Adam's line becomes starker: just as the beauty of Ophelia makes her death that much more tragic. And I promise that to many (most even) inorganic chemists, atheist or otherwise, a well-drawn molecular orbital diagram is truly, undeniably, a beautiful thing.

What do truth and beauty have to do with God? Well, being a

---

[2] (This and many sketches to come were made by the author via an awesome chemistry program known as ChemDraw.)

Hillsdale College graduate where the motto is *Virtus Tentamine Gaudet* but it might as well be 'Truth, Beauty, and Western Civ.', I see all scholarly pursuits as the pursuit of the true and the beautiful, including, of course, inorganic chemistry. But what is the most beautiful thing in all of history; both the past and what will be history in the future? It is a man tortured to death on a cross; because that man was God in flesh who died a filthy sinner, having taken our sins. The most beautiful thing is God's love and compassion, found at Golgotha on the cross.

I believe that this all ties together in the answer to one simple question: 'Why did God choose to not reveal the truth about what we call 'discovered truth'?' There are a whole host of medicines and natural understanding that God could have told us directly, but did not. Why did God not tell us about sanitation, and vaccines? I have a three part answer.

Firstly, God left us the search for truth and beauty because the search for these things, among the unbelievers, is the unwitting search for God. Because what is truly true, and actually beautiful (like a molecular orbital diagram for $Cu(H_2O)_6^{2+}$), bears the imprint of God. A man who spends his life diverting himself and seeking distractions will find noise and hell. A man who searches for the truth of a matter may stumble and find God.

Secondly, the answer for Christians is found in Ephesians 2:10 "For we are his workmanship, created in Christ Jesus for good works, which God prepared beforehand, that we should walk in them." God does not despise our weakness: rather he permits us a dignity of action by leaving us work that needs to be done. He does not do everything for us, because he loves us and does not treat us with pride, rather, in a great display of Divine humility, gives us meaningful, critical, life-and-death work to do; work that must be done; work that won't be done if we don't do it.

Lastly, God made it so that there were things to be found, unknown knowledge that could be sought out. "It is the glory of God to conceal things, but the glory of kings is to search things out." (Pv 25:2) There is a joy in tackling a problem and a glory in discovery. These too, are God's gifts. He has given us a great game of science to play, and we play it best when, like little children, we fail and fail and fail and then succeed and prance around the house in joy. (I have a toddler...)

So, I hope throughout these essays, perhaps while talking about them with others, you can see science as I see it: a tool fashioned by Christians. It has largely been used against the church in these recent years. However, I suspect that this is due to the abandonment of the sciences by those of faith. In the past century at least, the faithful have fled music, art, science... for an invisible cloister, so is it any wonder that the arts and the sciences seem to be our foes? No, however, these are just tools that Christians dropped, that have been picked up and repurposed by the world. We are entering a time where the Church militant cannot afford to leave any of its weaponry on the ground, we will have to pick up our respective weapons and fight, or be driven back into the catacombs.

Study Work

1. Define truth
2. Define beauty
3. How do these apply to your vocation?
4. Is truth always beautiful, and beauty always truthful?

Essay 2: Oda Nobunaga's Acceleration Due to Gravity

Science is the study of the world around us: the world we can see, observe, experiment on, touch, and (for the very brave souls) taste. Therefore, true science depends upon a particular epistemology: that is, philosophy of knowing. Depending on how cultures answer that age old question, 'How can you know?' will change how science is done, and how much progress at understanding the natural world science can make.

In the Lutheran church today, we confess with the ancient Christian faith that 'I believe in God the Father Almighty, Maker of heaven and earth.' From this we teach and confess that God, our trustworthy gracious God, made the heavens and the earth. This imparts a particular epistemology to the Christian mind. We can know that our observations of creation are valid, in the sense that the same experiment, done the same way, will yield the same result for any person at anytime, anywhere. We gain this confidence because our God is not one who lies, and is not one who despises material things – the incarnation itself puts that to rest. This confidence in the communicability of knowledge is a solid foundation for science to rest upon.

Most modern science is descended directly in philosophical belief from what might be called a Cartesian[3] understanding of truth. This understanding is 'modernist' and does not require that one believes in a personal God (many scientists of the 'modernist era' were Deist, not Christian). However, the epistemology still results in the ability to communicate ideas to anyone of any culture. Today we have a different epistemology, postmodernism, rampant in our culture. I will illustrate what I mean by this and the effect this transition has had upon the sciences below.

There is a standard diagram in Chemistry which illustrates the energy of a reaction as the reaction happens. For all reactions,

---

[3] From Rene Descartes.

there is an energy barrier between the reactants and the products. A general form of this diagram is shown in Figure 2.

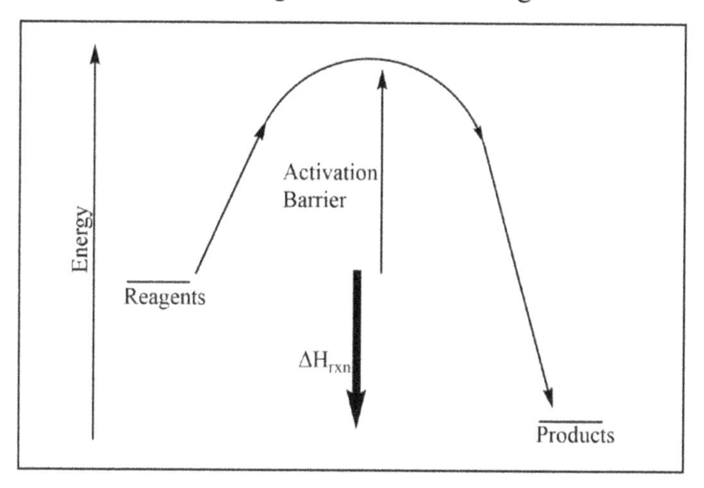

Figure 2: Generic reaction energy diagram

The reagents (reactants) start at a particular energy level. During the course of the reaction, they adopt structures which require an actual increase in energy. However, the overall reaction produces products which (in this case) are lower in energy than the reagents. Essentially, if the reagents can get to the peak of the 'Activation Barrier', they will make products.

The Descartes proof of the existence of everything starting from the famous 'I think, therefore I am' can follow a similar path (Figure 3). The only thing Descartes is sure of is that 'I Think'. It is less sure, but a logical step to believe that since 'I think' there must be an 'I' who thinks: 'I am'. However, this certainty is not enough to claim that what I observe is the same thing that you observe. So another step follows: 'I exist because Someone made me exist'. Indeed it is a derivative of the Ancient Christian 'Maker of Heaven and Earth'. Since God exists, then what we observe around us must exist at least generally, how we observe it. Notice that again we have reached an epistemology that is fertile ground for Science. There is no such thing as a different science for Chinese than for Armenians. If the acceleration due to gravity is 9.8 $m*s^{-2}$ for an Englishman in the 21st century, it must have been (also measured at sea level) 9.8 $m*s^{-2}$ for Oda Nobunaga in 16th century Japan as well as for Nebuchadnezzar. This is to say that tested, tried, and replicated results are objective, and not dependent on who is in the room observing or the culture they come from or the language they speak.

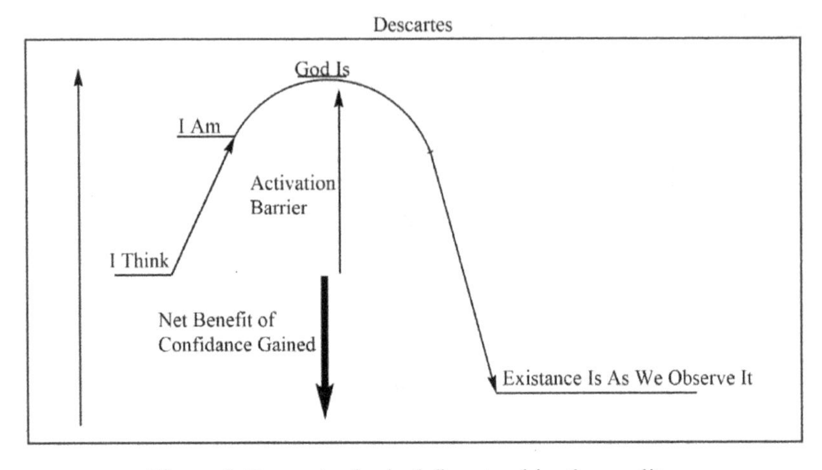

Figure 3: Descartes logical flow to objective reality.

The post-modern 'proof' of existence is similar to Descartes, but stripped of anything ultimate (Figure 4). A post-modern epistemology is entirely subjective: both truth and beauty is in the eye of the beholder. The postmodernist observes himself thinking: and then thinks about the fact that he is thinking, and thinks that maybe he might exist in some reality – something that might be more than himself. The result is that for the postmodern epistomology, there is something out there, but it exists largly (or entirely) as a construct in our own minds. Everthything is subjective, including ideas, thoughts, truth, good, evil, right, wrong, cowardice, and valor. Since existence is defined by the mind and thoughts of the one existing, it is not necessarily the case that something observed by one is true to the other. Oda Nobunaga's reality might have had an acceleration of gravity that was 5.6 $m*s^{-2}$ (which might explain ninjas…).

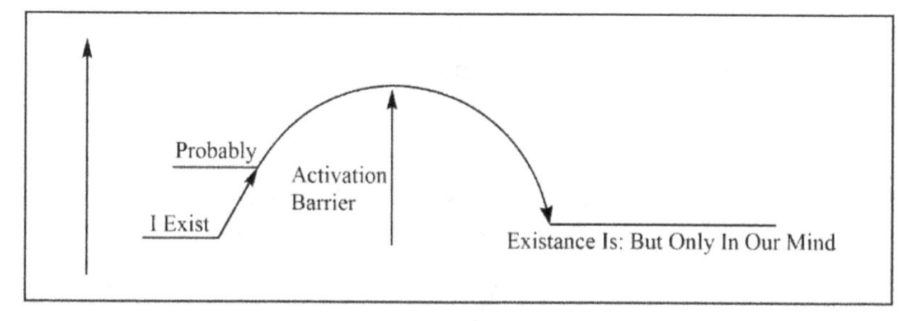

Figure 4: Postmodern logical flow to subjective reality.

If I, for example, observe a phenomenon, generate a hypothesis, and then test it repeatedly showing it accurate; for a postmodern, that too, would merely be 'my' truth, and not necessarily true for him too. This worldview is antithetical in every way to science. It is also antithetical to Christianity, a religion dependent on revealed and objective Truth.

Here we see why it is that, as is the thesis of every essay in this book, Science (qua science, not, indeed most scientists) and Christianity are not only compatible, but also fighting on the same side of a great civilizational war against between objective truth and postmodernism. Scientists, even, perhaps especially, atheist scientists, ought to see that their discipline is the next casualty of the culture wars and ought to recognize that making some common cultural cause with the strict traditionalists of the Church in the arena of absolute, knowable, transmittable and immutable truth is beneficial to the long-term health of science. For if truth is only a construct of the mind, then science cannot exist. Science depends upon the core belief that the truth is discoverable and objective.

Indeed, the first assaults on the fundamental epistemology of science are occurring now. Imagine if truth is not timeless and knowable, how could one know how to operate? The common answer from the postmodern world seems to be that a consensus of quality minds actually defines truth and therefore science. We see this in the standard defense of anthropogenic global warming. (It is not my purpose to discuss the evidence for that issue here, only the standard defense of it.) We are incessantly told that 97% of experts agree that global warming is caused by mankind and will be catastrophic in its effects. This argument is postmodernist in nature. There is no appeal to evidence, experiments, data, -- there is only the consensus of experts. However, science does not bow to the consensus of expert opinion, the world does not change to match the conclusions experts have drawn about it. Otherwise, we must believe that the sun did *indeed* revolve around the earth for Ptolemy, while later, after some cataclysmic solar revolution, it now is orbited by the earth due in large part to Galileo. This ludicrous conclusion is the logical outflow of the postmodern, God-less, epistemology.

Every scientist, and probably every person not fully committed to the conclusions of the postmodern epistemology, would find this ridiculous. And yet one academic department after another has fallen to this worldview. Perhaps because they are not grounded in observation, or perhaps because scientists are a particular brand of curmudgeon, the humanities have fallen first, long ago. (Think: 'Beauty is in the eye of the beholder.) Now however, even the hard sciences are under attack. The global warming consensus is (although it mayn't be a consensus after all) the only consistent argument in favor of the theory and people peddle it as fact. Man-made global warming may indeed be fact, but the only evidence presented is that 'all the experts agree' and that 'all the expert's models agree' which statements are, under the surface, identical. Now Truth is in the eye of a consensus of beholders.

Having investigated briefly the effect of the postmodern epistemology on the sciences, we can look now at what it does to religion. If truth is only a construct of the mind, objective religion cannot exist. Religion is demoted to 'something that makes you feel better' which puts it into the category of whiskey. If that is the case, then anything in the religion that makes you feel upset or threatened must not be true –for you. Maybe someone else feels better knowing that God condemns sexual immorality of all sorts, then this religious belief is true for them. And if a large enough group of *spiritual* people agree that something is acceptable for god, then, it must be.

That of course leads to the absurd idea that god was pleased by the Mayan human sacrifices. If that were the case, then you could very easily argue not just that societies make their own gods in the image that pleases them, but that god must also *actually* exist for their believing created it. (If you read *American Gods* by Neil Gaiman you can see this work out in fiction. You can also read *Hogfather* by Terry Pratchett if you want the same kind of vision with less apparent nihilism and violence.) I am sure this scenario sounds more plausible than the Ptolemy scenario above. That missing gut 'that's dumb' is just due to the fact that you, dear reader, live in the world (though hopefully are not part of it) and you breathe every day the subjectiveness of the world that has ruined every domain but the hard sciences, and is rapidly encroaching on that final redoubt of human thought.

So that is why I recommend not just a peace treaty between traditional Church leaders and scientists but active cultural cooperation. The Church is not in any danger of being wiped out by a fleeting theory in the minds of men, but science may well have to come and take shelter once again inside the doors of the Church, running for sanctuary from those that scientists currently think are science's friend. This is also why Christians ought not feel as though the study or practice of science is unhealthy for their souls. In fact, the study of science and the pedantic drive most scientists have for diamond hard objective truth is a healthy mindset for every Christian.

Study work

1. Many things are considered 'true'. What 'true' things vary from place to place and time to time? Are these things truth?

2. In many times, in fact, perhaps in most times, human sacrifice was (is) considered not just acceptable, but honorable. In some cases, the families of those sacrificed would be honored by the selection of their family member. Can you be sure that this practice is wrong always and everywhere? Defend your answer.

3. Develop two separate arguments that abortion is wrong no matter how early. One argument must be based in morality and Scripture, the other entirely based in science and logic. What do you think the benefits of having entirely separate arguments are?

Essay 3: The Probability Density of Truth

It is a question for all ages: What is truth? What is *the* truth? What is the truth about *this*? The search for truth is one of the defining characteristics of western culture. (By that I mean the many cultures that have grown out of the Christian worldview, generally also influenced by Ancient Greece and Rome.) The belief that truth is at the very least *theoretically* knowable or discoverable is the driving force behind the science, the philosophy, and the art of western culture.

In these days of all-out assaults on this culture, the desire to know truth, the belief that it even exists objectively, is out of style. In many ways, today is similar to A.D. 410 when the Vandals sacked Rome, or perhaps A.D. 732, just before the Muslim conquest of the West was stopped by Charles Martel. Only now the danger comes from within the minds of the western elite themselves. "Pride and a little scratching pen"[4] have been used to end the search for truth and the belief in the Truth that makes the west itself. And so, we should talk about truth.

Obviously, in one little essay, no one could actually answer that question. So, what I will do here is discuss some of Truth's characteristics. What kind of thing is truth? There are three notions about the nature of Truth that I think are fairly common. (I suspect the first two are much more commonly held than the third.)

1. Truth is relative to every individual, hence individuals 'create' their own truth.
2. Truth is always found as a compromise between extremes.
3. Truth is the single point that is not falsehood.

The first definition is worthless, by which I mean: that definition is useless. If a word only means something to every individual person, than that word (and hence the idea it reflects) might as well not exist. Words are reflections of ideas by which

[4] Chesterton, G. K. *The Ballad of the White Horse.*

we attempt to communicate with other people. And while it is true that to some extent every word and idea's meaning is colored by the person understanding it, that emphatically does not imply that it only has meaning insofar as each individual wants it to. The very fact that I am describing some aspects of what I believe the Truth to be means that I also believe that ideas about truth are knowable and communicable; something that anyone who believes #1 and tries to convince other people about it must also believe. Hence, someone who believes that they create their own truth, has the same logical problem as the person who believes that there is no absolute truth. The former has decided that 'their' truth is that they make 'their' truth, and then they promptly attempt to convince everyone that their truth is *the* truth.

The second idea about truth, that it is always a compromise between extremes is due to a misunderstanding of the golden mean, and what compromise really is and what it is really for. For instance, Bill Clinton either had sex with that woman or he did not. The truth is one or the other, and a compromise between the two extremes is laughable. Compromise exists because while principals and ideas can be narrowed down to logically sound statements, how to apply those principals in this messy, messy, world we live in is very confusing. Two people with the exact same set of principles (ideas about truth) can come to two very different methods of implementation. And hence, if we have to work together to make something happen, we may have to compromise our methods even if we are in a situation where none of our principals are in conflict.

I do not think that the Truth, qua Truth, is any different in nature than the truth about an event. The nature of Truth remains the same, the difference is scale and applicability. So this leaves the last idea, #3. To envision what I mean, do not think about truth as a line or a continuum, but rather as a Gaussian distribution. When you first encounter some aspect of the truth, your understanding is like the dotted line in Figure 5, perhaps centered correctly, but vague and spread out. As you learn more about the Truth, your understanding grows sharper and more accurate, like the solid line in Figure 5.

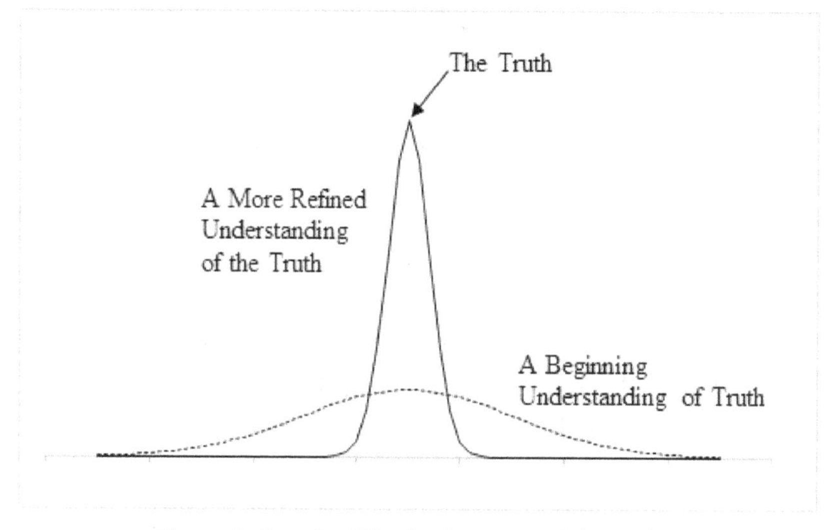

Figure 5: Gaussian Distributions around the truth.

However, the Truth is that point at the very top. At this point, movement in any direction rapidly becomes false, or a lie. (Incidentally, I think that slope down is the infamous 'slippery slope'.) That is why from the bottom (e.g. from the viewpoint of a lie) the truth is huge, imposing and terrifying, and from the top, the truth is tiny, fragile and… terrifying. The Truth itself is something too good to not be frightening to sinful man, and yet the attempt to stand on its pinnacle and believe the truth, and of course also believe in the Truth as embodied in Jesus Christ, is the heart of Western culture and civilization. At the risk of sounding like the sadly insipid Shepherd Book[5] who dies begging the main character 'to believe' but doesn't seem to care about what the belief is, I will say that anyone searching for the real truth, the knowable solid and sometimes terrifying truth; is close to the heart of the West irrespective of what they actually believe. In this way, the search for truth is a path that God has made to draw people to His word, through which they may be saved.

Being Christian myself, I believe that I have a Book that directly reveals as much of the Truth as man's mind can handle

---

[5] If you don't know who this is, you should to watch the TV series 'Firefly' and the movie 'Serenity'.

from the Creator of the universe and our minds. In this I find the Truth, the whole Truth and nothing but the Truth. It is the comfort, the strength, the sword and shield and song of what we crassly call Western civilization, and could perhaps be better called the civilization of truth-seekers.

Study work:

1. Draw a diagram representing Truth and the various attributes (such as ugly, beautiful) or areas of human though (such as art, science, economics) etc. Discuss your diagram with others (if possible).

Essay 4:
Battle Lines of the Church Militant

There is a crisis in __________. You fill in the blank. From a Christian perspective the list of crises is immense: the arts, education, the church, the middle east, the world, college education, science, culture, politics. Any one of those and many more could easily be a topic of whole books. Many of them are. It is easy to become overwhelmed by the onslaught. Just when the church is moving against scientism, modernism, and evolution; the whole culture slips sideways and attacks from a post-modern, subjective stance where an XY chromosome pair is not always male and having ovaries doesn't even *strongly imply* that one is female.

In this situation, it is important to remember that the Christian faithful on earth are part of the Church Militant. This isn't the 'church somewhat angry': we use the word militant in common usage somewhat sloppily. This is, in fact, the church in pitched battle. And as with any campaign, each battle has a different landscape, a different arraignment and quantity of enemies and so forth. If we consider all the crises facing the church today, we will find that these are not *individual* crises, but different positions on the battle lines of today's Church Militant. The bad news is that we are surrounded entirely.

At the Battle of Ia Drang, shown in the best war movie ever, 'We Were Soldiers', the American forces were entirely surrounded. Figure 6 shows the battle lines on day two of the battle at Landing Zone X-ray.

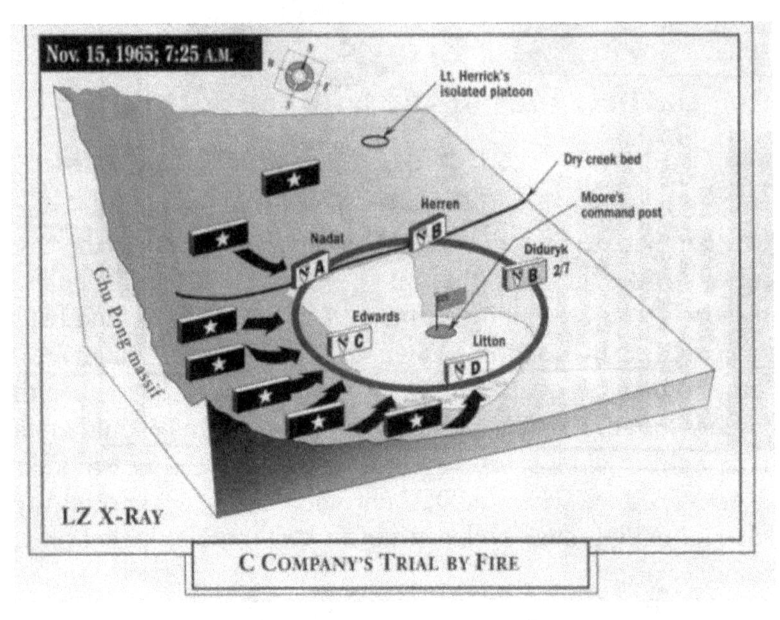

C COMPANY'S TRIAL BY FIRE

Figure 6: LZ X-ray, Day 2.[6]

You see how they were surrounded… in a valley? That, I believe is much like our present position as the church militant. Everyone needs to be armed for this battle, and everyone needs to find their part of the battle line and stand their ground.

I guess that a fair number of readers will believe that the sciences are one of the enemy units surrounding the Church. This is the attitude of quite a large number of Christians that I have talked with. The scientist is automatically suspect, but the youth pastor with the bad doctrine is totes ok. While bad doctrine in the church is itself a major issue, it is neither the focus of this work, nor my area. I am a scientist.

To go on, we need to answer the difficult questions: what is science? And what is a scientist? Science is not automatically believing an authority, or even a consensus of authorities. Science is like a toolbox. It is full of mental tools like calculus, molecular orbital theory, and matrix multiplication, as well as physical tools like Schlenk lines, gas chromatography, and x-ray diffractometers

[6] Galloway, Harold G. and Moore Joseph L., *We Were Soldiers Once . . . And Young - Ia Drang, The Battle That Changed The War In Vietnam.*

(to name tools that I have used/ am using). Science then takes observations and brings together other understanding to answer questions about the physical world using extremely hardheaded and pedantic logic. In short, science is a tool or to return to our church militant metaphor, a weapon. It is, in fact, a weapon forged by Christians. It also happens to be a weapon that we have abandoned on the battlefield. How can we be surprised by the fact that others pick up this weapon and wield it for their own ends?

So what is a scientist? While technically credentials do not make a scientist, it is difficult to achieve the necessary training in mathematics and scientific theory and well as practice investigations without at the same time accomplishing credentials. This will sound elitist (and I loathe elitists) but it is intended to convey a truth. It is hard for a non-scientist to envision the effort required to actually *become* a scientist, let alone actually *be* one. Therefore, people underestimate the effort required, and so are able to assign a lot of motives to scientists that largely do not exist. For something as difficult as becoming, say, (not to toot my own horn too much) an Inorganic Chemist, there are only a very few motivations that will suffice.

Most scientists (who nearly all believe evolution) are *not* directly motivated by 'hating God' or by guilty consciences for some secret sin. Most simply scientists *enjoy* thinking about their field. They are either more nor less opposed to God than any unbeliever. Often Christians think of the Atheist Scientist as anti-God shock troops, like they are the devil's Green Berets. This is simply not the case. While scientists are as corruptible, venal, and sinful as your atheist plumber, they are also frequently conscientious, honest, and hardworking (again like your atheist plumber).

Here is the truth: most scientists believe in evolution because *other* scientists tell them that the theory is the only one that makes sense. Everyone believes nearly everything based on someone's authority. You probably know that 2 + 2 = 4 because you can work it out with fingers, but you don't know that there are such things as atoms and electrons by your own observation. You take my word for it (I'll gladly stand in for chemists everywhere…). In the same way, one scientist takes another's word for theories. In fact, the only direct evidence that I, personally, have collected that atomic scale particles exist is X-ray diffraction, and the reason I believe that it tells me what it tells me is only because I accept that some other scientists are correct about waveform interference. All of scientific knowledge is like a pile of stones, where we add on new stones upon the older rocks. So it is entirely natural that most scientists believe what other scientists tell them about evolution.

(A brief diversion to a good question for introspection: How do you choose whom you will believe?)

We could divide scientists (and really, everyone) into four groups, which conveniently can be plotted on an X,Y coordinate system. In this case, the X axis is 'Honesty' where the left side is negative (negative honesty values would be…bad.). The Y axis is 'Correctness', and negative correctness would be, incorrect. As you can see in Figure 7, there are correct honest people (quadrant B), incorrect honest people (quadrant D), and correct and incorrect charlatans and grifters (A and C respectively). This applies to the scientists as well.

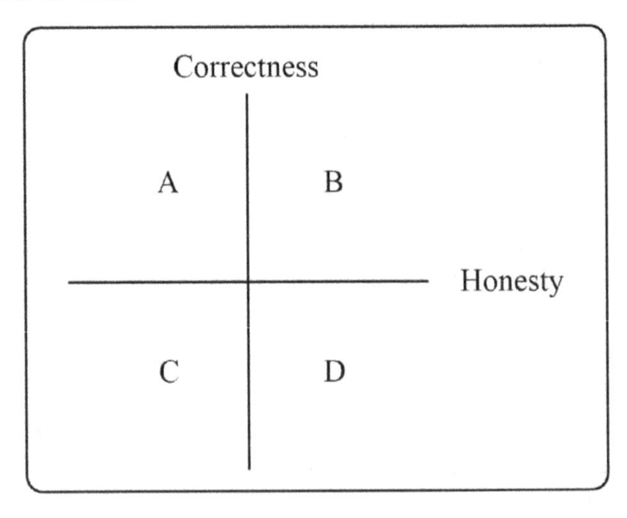

Figure 7: A plot of possible combinations. (Remember, just because the quadrants are equal sizes doesn't mean that measured data would fall equally into each quadrant.)

So, a great many scientists, perhaps even most, are with respect to evolution honest and wrong. Likewise there are some creationists who are right but are using the fact that many evangelicals agree with them as an integral part of their con. I think that viewing science and scientists through this lens is evidence enough that the sciences are, or ought to be, a tool in the hands of Christians, not considered enemy territory.

A good example of a place where scientists are sorely needed in is the modern transgender insanity (see the final essay in this volume), having Christians whose vocations are in the sciences would be a great help in defending against that onslaught, which will, in fact, be more brutal than that of evolution a century ago. Another example of science being on our side is that of the abortion debate. Through the hard work of Pro-life advocates, much good work has been done. But the biggest tool they have had to change the culture came out of the sciences: the awesome improvements in ultrasound technology. It is indisputable that the cultural mindset has been more swayed by excited parents showing ultrasounds in which you can see a cute little baby than by every opinion paper or editorial ever written. Science can afford many such tools, we as Christians need to go and get them.

So, is there a crisis in the sciences: indeed yes. It is quite bad, especially the less 'mathy' you get. Iceland is has nearly exterminated all its Down Syndrome children via science.[7] "More than 70% of researchers have tried and failed to reproduce another scientist's experiments, and more than half have failed to reproduce their own experiments."[8] Many commenters on climate science seem to think that consensus is what determines truth.[9] As Chesterton writes: 'The sky grows darker yet, and the sea rises higher.' But this simply is an argument for young Christians (who can do math) to re-enter the sciences, and to do so aware of, and prepared for, the specifics of their part of the battle lines of the Church Militant.

---

[7] https://www.cbsnews.com/news/down-syndrome-iceland/

[8] Baker, Monya, *Nature*, Vol 533, 2016.

[9] Legates, D.R., Soon, W., Briggs, W.M. et al. Sci & Educ (2015) 24: 299.

Study work

1. Think of several fronts in what we are calling the current battle between the Church Militant and the 'Devil, the World, and our flesh'.

    a. Think of what personality types and what vocations might be best suited to which areas of the battle.

    b. Think about your place, your part of the line.
    c. Make a diagram of your thoughts
    d. Discern the assaults other people at other points in the line might be vulnerable to, so that you can encourage your Christian brothers and sisters in their faith and in the good work that God has given them to do.

Interlude

The Ballad of the White Horse
Book 1: The Vision of the King (G. K. Chesterton)

"The men of the East may spell the stars,
And times and triumphs mark,
But the men signed of the cross of Christ
Go gaily in the dark.

"The men of the East may search the scrolls
For sure fates and fame,
But the men that drink the blood of God
Go singing to their shame.

"The wise men know what wicked things
Are written on the sky,
They trim sad lamps, they touch sad strings,
Hearing the heavy purple wings,
Where the forgotten seraph kings
Still plot how God shall die.

"The wise men know all evil things
Under the twisted trees,
Where the perverse in pleasure pine
And men are weary of green wine
And sick of crimson seas.

"But you and all the kind of Christ
Are ignorant and brave,
And you have wars you hardly win
And souls you hardly save.

"I tell you naught for your comfort,
Yea, naught for your desire,
Save that the sky grows darker yet
And the sea rises higher.

"Night shall be thrice night over you,
And heaven an iron cope.
Do you have joy without a cause,
Yea, faith without a hope?"

Essay 5:
Wise as Serpents

When I was in college, I had a great roommate. I was a Chemistry major and he was a Classics major. I don't remember the exact context, but I was giving him a hard time about something he believed after reading... on the Internet, no less!

I distinctly remember telling him that while he was as gentle as a dove, I was wise as a serpent – and the Internet was no place for doves. While this is a rather crude approximation, it does demonstrate one of the vocations of the scientific mind. Scientists are usually skeptical, literal, and almost always pedantic.

It is important to note that scientists are not like this because they have spent their lives in the sciences. It is more accurate to say that they spend their lives in the sciences because they were made this way, by God. My roommate, on the other hand was trusting, metaphorical, and genial.

Both the scientific and artistic-literary types belong in the Body of Christ. Both types have an important role. The second type is easier to get along with and "comfortable" in any setting. Not so, the scientist. Thus, the importance of the pedantic type is often overlooked.

Keeping the mental nature of most (if not all) scientists in mind, we can look at the relationship (and some would claim, conflict) between faith and science. Talking about 'faith and science' is superficially easy. In fact, there no sentient entity of 'faith' nor is there one for 'science', and so it is more helpful to discuss a scientist in a church.

The first issue is this: we do not have to surrender the sciences because they are inherently hostile: they are hostile because we have surrendered them! Christians who are called to a scientific vocation can be viewed as something like the first squad landing on Omaha beach. The sciences used to be dominated by Christians and at some point Christians seem to have fled like the British at Dunkirk.

The young scientist of Christian Faith entering the battle must be prepared to be mocked and ridiculed. If he tries to publish something that has even a whisper of dissent from the lockstep atheistic shibboleths (such as evolution) he will not be published. This is why it is important to find a way to build up the scientist's faith–to find a way to help treat their battle wounds.

The critical question is, how can the church encourage and train Christians to be both faithful believers and excellent scientists. It is one reason that I teach chemistry at CUW in Mequon, WI-- to participate in what I believe is a part of the answer. Another part of the answer is well illustrated by the witness of St. Thomas.

In this testimony, we find that demanding evidence, being skeptical, and being hard-headed is allowed. (Hard-hearted, Thomas was not. John 11:16; "So Thomas, called the Twin, said to his fellow disciples, 'Let us also go, that we may die with him.'") Also, when viewed without the factually correct but emotionally misleading 'doubting' tag on Thomas, the account can show the role of the scientist in the church. Here we find a salve of encouragement for those whose vocation leads them into constant conflict with the world on questions of evolution, the age of the earth, and the feasibility of miracles.

It is important to note that none of the disciples believed without evidence. From their previous actions, it is reasonable to think that most (or all) of the disciples would not have believed if they had been in Thomas' position. Thomas happened to be the one not there. But the doubt of St. Thomas is really something solid that a skeptic, a born doubter, a scientist, can hang on to.

St. Thomas proclaimed that he would never believe in the resurrection, not unless he gathered his own evidence. The evidence required: 'Unless I see in his hands the mark of the nails, and place my finger into the mark of the nails, and place my hand into his side, I will never believe.' is evidence for both the death and the resurrection of Jesus Christ. It is the evidence for the very core of our faith; imagine what it would be like if the evidence had not been asked for. Then, Jesus in his great mercy, both for St. Thomas and for us now, came again and provided the evidence. 'Then he said to Thomas, "Put your finger here, and see my hands; and put out your hand, and place it in my side. Do not disbelieve, but believe.'"

The evidence here given is accurately categorized as empirical evidence. The empirical evidence that St. Thomas collected is recorded by St. John in a document that has a high quality textual transmission. (The fact that it is a religious document does not negate the excellent textual transmission of the writing of St. John to what we read when we open the Bible.) In effect, this account places you three steps away from empirical evidence for the Risen Christ (You → Textual Transmission → St. John → Thomas putting his hand in Jesus' side).

I recently thought through my 'distance' from the empirical evidence that electrical charge is quantized. This is foundational knowledge for chemistry, one electron's worth of charge is the smallest amount that can be transferred at any given time. There is no way to transfer 0.33 or 1.5 electrons worth of charge; the smallest charge is one electron exactly. The empirical evidence for this is a classic experiment known as the Milliken Oil Drop experiment.[10] Interestingly, the chain of evidence between me and empirical evidence from the Milliken Oil Drop is something like Me → My College physics professor → Textual Transmission → Milliken's empirical evidence. The 'distance' measured by steps in the chain between me and a fundamental physics concept from the early 1900's is the same as that between me (and you) and the Resurrection. I have tried to summarize this idea in Figure 8.

[10] It is fascinating, you should look it up, but it is beyond the scope of these essays. It is a bit too technical.

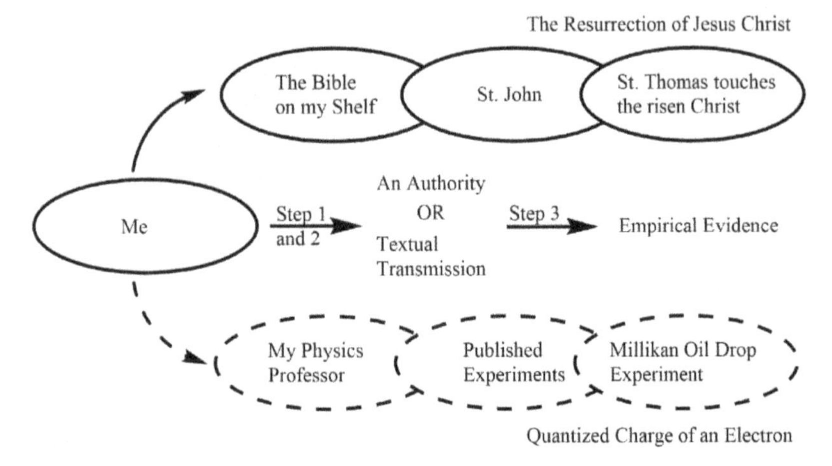

Figure 8: A 'Chain' of Transmission for empirical evidence for two entirely different events, with the same kind of knowing involved.

This illustrates that the scientist in the church does not have to fear hard-headed inquiry or examination by skeptical outside non-Christian observers. We have the reassurance that, since what the Scriptures teach is true, investigation and analysis will show its truth. We also see that the role of the scientist in the church includes being the one who asks for verification: for little things like something on the Internet and for big gigantic things like asking for evidence of the death and resurrection of Christ himself.

Lastly, and most importantly, we find in the account of St. Thomas the most meaningful part of the resurrection account to a skeptical mind: 'Thomas answered him, "My Lord and my God!"'. The witness of St. Thomas is the witness of a skeptic, of a man wracked by doubt, a man who almost certainly awoke, lived with, and went to bed accompanied by doubt. The hard-headed skeptical witness of St. Thomas to the death and resurrection of Jesus is evidence that a scientist can believe. And if God can raise Jesus from the dead – there is nothing, not creation, not miracles, that He cannot do.

I think as a final point, from a scientist to a largely non-scientist community, it is good to remember that everyone has their own cross and their own doubts. A theologian can, and ought, to struggle over why some are saved and others not, and it would be our calling to help and lovingly assist our brethren in their trials.

Likewise, it is our vocation to help those who have doubts born of scientific knowledge carry those doubts and to point to Christ and Him crucified first and foremost. And perhaps, while pointing to the Risen Lord, to also point to the witness of St. Thomas.

Study Work

1) Make a flowchart indicating each step between you and empirical evidence that Baron von Steuban trained the continental army at Valley Forge in the American Revolution. Do you believe that he was there? How different is that from the empirical evidence for the resurrection given by St. Thomas?

2) Pick any of the following (at least somewhat) controversial science related topics. Consider what you believe about them and the evidence you have for that belief. For one that has empirical evidence, make a similar flowchart for as in the first question. Now (for the important part) evaluate each link in the chain and decide if they are trustworthy links.

Fetal Stem cell research
Umbilical stem cell research
Gene editing (CRISPR)
Anthropogenic Global Warming
GMO foods

Essay 6: The Self-Flipping Light Switch

Imagine a dedicated and precise scientist. He grew up friends with Tycho Brahe and has mustaches to compete with Brahe or even Hercule Poirot. He is an expert mathematician (we could call him… Bycho Trahe). And then, by some freak accident he is catapulted into the modern world, and so obviously has no understanding of lights and light switches. He observes the same panel of three switches repeatedly and takes careful notes on some old timey paper he has with him. (A scholar is never without pen and paper.)[11] He controls every variable he can think of, so he observes the switches at the same time every day, 11:00 am for twenty days. He observes for many days and finds out that sometimes the switches are up and sometimes down. He also determines after the second day that the switches and the lights are related to each other. He calculates the time per flip of the switch. Sometimes it is several days and sometimes it is changed the next day. On average, each light switch flips every 1.58 days. (And the average is fairly similar for all three switches.) He then notices from his data that, since there are three switches, at least one of them is flipped from the previous day every single observation. Then, being a committed scientist, he sets out to see a switch flip. He figures that even if the unusual happens he will only wait a day or maybe two before he sees one of the switches flip. He arrives and all three are down, and he sits down to observe. He waits, confident in his assessment of the probabilities. He knows that at some point, a switch must flip. There are three switches and the time between flips is 1.58 days per switch, so it should happen any time. He waits a week, and gets very hungry. He hadn't planned for this much watching, they should have flipped many times already. After another week, he was ready to give up, when for the first time, the owner of the house walks in, tanned from vacation in Florida, and turns the light on at 1am, waking up a terrified

[11] Lewis, C.S., *Prince Caspian.*

Bycho Trahe and freaking out a bit himself until the mustaches defused the situation.

What was Bycho's mistake? Despite all the probability calculations he might have done, despite all the effort, since he could not begin to comprehend what was actually happening, no amount of probability calculations could help him. He either did not know, or was unwilling to admit, that the light switches were not switching themselves. He was observing something without taking into account the cause of that something. He had calculated a probability for something that was not possible.

I hope that with this anecdote (certainly not believable) I have outlined a vision for how one might, with pure intentions, calculate the probability for something that is not possible.

I suppose this trend began legitimately with quantum physics, where, for example, we calculate the probability that an electron will be in a particular area and call it an atomic orbital. In science there are many shades of scientific rigor between one field and another. While it is fun to jab at biologists for not being 'real' scientists… as a chemist I'm vulnerable from the physics flank… and they, perhaps fall to an argument with the mathematicians. The classic 'who's a real scientist?' argument is amusing, but it is fit really only for a spoof article. Despite this, it is undeniable that respected (and respectable) institutions, such as the sciences, accrete to themselves grifters, hucksters, and charlatans. And one common trend is an overabundance of probability calculations standing in for logic.

One cannot say 'Every biologist is a hack' (though it would amuse me to pretend it…) but it is true that any biologist might be a hack. A chemist might be incredibly gifted, well trained, and even well-intentioned, but selling a fraud. How is this possible? Well, it may surprise some, but even the 'mathiest' scientists are people too. It takes a great amount of introspection to begin to see one's own preconceptions. It is invariably the case that a scientist will collect much information and will have to make the choice as to what information is the important information and what is the result of improper techniques. What might seem to be important information to you seems so obviously irrelevant to someone else that it gets, honestly, genuinely, ignored.

When scientists and their mimics get really going, there is one implicit assumption made quite frequently; honest and dishonest alike. The argument actually is never explicitly made. It shows up in the following types of context: 'A probability has been mathematically calculated with rigorous observations and taking every variable into account. The probability is very small, but given enough time it will have to have happened many thousands of times.' You feel like this makes sense. If we found out that 0.0001% of the population is a telekinetic, then we would have every right to expect, due to the billions of people, some thousands of telekinetics on the earth. However, this line of thinking is not automatically true, it is true under one condition. That is this: whatever the probability is predicting must *first* be possible. The possible is a binary situation: either something is possible or something is not possible. Either a telekinetic can be born or one cannot be born. No amount of probability calculations will make a telekinetic be born if it is *not possible*. You might argue that clearly, telekinetics are not possible, but scientists are only calculating things that have been observed and therefore are already proven to be possible.

Let's think about evolution for a while. There are several fundamentally distinct steps in the necessary timeline for random chance to product intelligent life. In Figure 9, I have summarized some of them, although one could break down those steps even further or add numerous events to this figure; the main point is generally applicable. (I have put the chemistry topics, those actually near my area of expertise, in italics as they will serve as examples.) It is most important to remember that this list of necessary, accidental, items is nowhere near exhaustive, it is merely illustrative.

A) Physics events

Big Bang

Formation of matter
Formation of stars
Formation of planets
Etc.

B) Chemistry events

Atoms to small molecules

Amino acid formation

Amino acids to proteins

RNA formation

RNA Transcription enzyme formation

Lipid bilayer formation

Etc.

C) Biology/ biochemsitry events

Non-living chemicals become living chemicals

Macro evolution of the first living cell
to multicellular organisms via addition of
external randomly formed genetic code.

Development of intellegence

Intellegent Life

Figure 9: List of selected evolutionary events critical to occur, randomly, in order to form living cells. (Some events could predate other events, or happen at the same time, in theory. Some of the listed events are proven to happen randomly: e.g. Atoms to small molecules.)

For each of these events, the probability of it happening is calculable. Certainly, the scientist must make some reasonable assumptions, but if you fault him for doing that, you must also fault yourself to the reasonable assumptions you make every day. This is not where the problem lies. Let us consider the 'amino acids to proteins' step. There are 20 biologically relevant amino acids. (A discussion for another time involves the fact that all 20 require the same chirality to be biologically relevant.) Proteins generally contain hundreds of amino acids. So, it is entirely feasible to say, 'if there are 20 amino acids and 350 amino acids, there is X chance that a random amino acid scramble with produce the exact right protein.' If you are aware of the rules of probability at all, you will see that this is a tiny chance. (I intentionally am omitting the number, as it has been my experience that almost everyone gets caught up in the numbers, the size of which they have no real grasp of, rather than the concept being discussed.)

While the chance is infinitesimal, the probability says that out of a very large number of accidental encounters, one will be the right thing. Here there is a logical flaw. The argument has been made so that the ability to calculate a probability stands in the stead of evidence for possibility. We are told 'if this has a probability, you must concede that with enough time it is guaranteed to happen.' However, if it is not actually possible (a completely binary proposition that is entirely unrelated of how common the event is) then there cannot be enough billions or even trillions of years to have the event happen. (Just like light switches do not switch themselves, at least not without a smart-home…) This has nothing to do with improbability.

You may point out that in the aforementioned calculation, we did not treat with the vastness of the so-called 'primordial soup' or include the odds that there would be any appreciable concentration of amino acids at any given point. This argument is really a fool's game if you are discussing with a scientist. Firstly, because controlling and predicting all the variables is really what scientists do (and therefore have likely already considered it). Secondly, no matter how small you whittle the probability down to, enough time will simply cure it. No matter how small the probability is, the equation remains: tiny probability + billions of years = it will happen sometimes.[12] Thus you ought to focus on the fact of possibility. In every chemist's experience, uncontrolled polymerization results, not in highly ordered structures, but brown goo.[13] And quite frankly, though unprovable (Who has a billion years on their hands, anyway?) it is my conviction, as a chemist, that every single uncontrolled reaction of amino acids will always, invariably, produce brown goo, and will never once make so much as a single α-helix. This illustration applies to many of the steps I listed under the 'B) Chemistry events' in Figure 1. If even *one* of those steps is actually impossible rather than extremely improbable, then evolution *cannot* have happened: evolution is then the theory of the self-flipping light switch.

Study Work

1) There is a classic paradox dealing with the apparent high probability of extraterrestrials and the lack of contact called the 'Fermi Paradox'. Read about the Fermi Paradox and then apply the probability/possibility distinction to the premises. (The conclusion is essentially 'So where are all the aliens?') Can this provide a valid solution to the Fermi paradox?

2) Think of a situation where the reasoning "Since there is a small but non-zero probability of X happening, it must happen relatively often if some large number of years Y passes" is accurate. What makes these true?

---

[12] In general, it is never a good idea for a scientific laymen to argue science-based apologetics with trained scientists. (Leave that to men like James Tour: http://www.jmtour.com/personal-topics/personal-statement/) The goal ought to always be to point to Christ and him Crucified and the historical evidence for the resurrection and take a pass on the science. However, when considering the impact of science upon your faith, and that of your friends, family and especially children, I hope that these discussions will be an invaluable tool.

[13] Tour, James M. Two Experiments in Abiogenesis, http://inference-review.com/article/two-experiments-in-abiogenesis

Essay 7: Choose Your Own Authority

We live in a world where everyone is worried about nuclear proliferation. What is slightly less noticed, but still a problem, is Authority Proliferation. In the world of the Internet there is an ever expanding number of 'experts' and the selection of authorities to trust is an important, difficult, issue. Christians are called to be both, and at the same time, cunning as serpents and gentle as doves (Matt 10:16). Many churches emphasize the latter, such that shrewdness is an underdeveloped trait. The result is that many Christians accept various people as authorities with little or no analysis of the credibility of that authority. There are several wrong ways to decide who to believe, such as: who you agree with, credentials, consensus of experts, and personal investigation. Each of these common methods has serious or fatal pitfalls.

It is normal for people to assume that someone they agree with is honest and someone they disagree with is dishonest. However strong this natural impulse of belief may be, it is the key element in any con artist's game. Every grifter counts on agreeing (or pretending to agree) with you on a lot of things in order to gain trust. Only then will the con-man strike. Dishonest philosophers, scientists, and theologians do the same thing. First, the grifter finds common ground, then he makes a shift to what the goal is: he strikes. Maybe the con-man wants money, maybe prestige, maybe for you to change some core convictions you hold. Invariably, the most dangerous fraud is the one that you agree with on a number of things already. In Essay 3, I briefly discussed the diagram in Figure 10. We briefly talked about the possibilities that scientists, and people in general could be simultaneously correct and a fraud or incorrect and honest. It is always better to seek out those who are honest, whether they disagree with you or not. Even if a fraud agrees (or pretends to agree) with you about the age of the earth, or evolution, or Justification, these types ought to be always avoided because the line between agreement and manipulation is vanishingly thin.

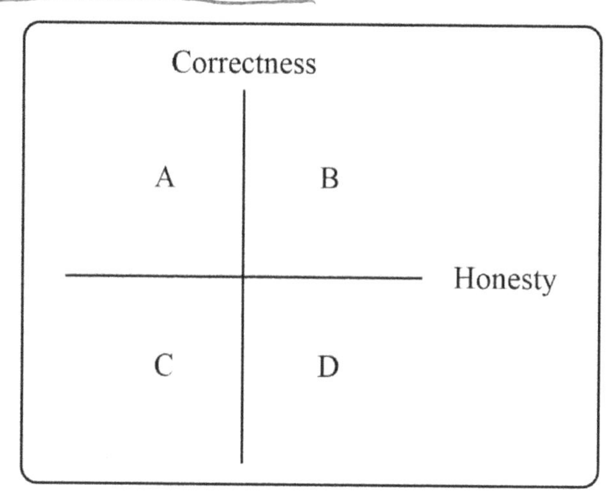

Figure 10: Reprise of Figure 7, Four Kinds of People.

So we can put away 'agrees with me' as a reason to accept an expert. What about credentials? While I believe that someone agreeing with you is *never* justification for or against accepting expertise, credentials can have some meaning. Not as much as those with credentials would like, but some impact nevertheless. For example, Bill Nye is billed as 'the science guy'. What, may we ask, are his credentials? He was trained as a mechanical engineer (not easy, lots of math, not exactly science) and has starred in kids shows as 'the science guy'. That about sums it up. So, his opinions about Global Warming, Human Population, Evolution, etc have little real weight. There is technically the possibility that someone like Bill Nye is a brilliant autodidact. Perhaps he has no real credentials because he learned everything about science by himself and knows so much there isn't any real reason that he should go back to school just to get credentialed. Or, perhaps, he is a gifted spokesman for a scientific/political point of view. His actual credentials support the latter strongly and do not give much credence to the former. The same dynamic is observed with what we might call theological credentials. Just as people pretend to know more science than they do, the often can pretend to know more theology than they do. While seminary training does not necessarily impart a love of doctrine and a commitment to teaching the Gospel in its truth and purity, not having that education, that credential, does indicate the lack of something substantial.

Unfortunately for credentials, there are plenty of highly credentialed people who are not much more than quacks and grifters. They peddle academically popular theories in order to extract large government grants to support their prestigious research life, which ultimately is nothing more than the academic's personal hobby-horse. The same can be true for pastors. There are plenty of seminary trained, ordained pastors who love nothing more than their bureaucratic prestige or give emotional doctrine-free sermonettes full of life experiences and devoid of Law and Gospel. So, while credentials can carry some weight, they really cannot be the deciding factor in trusting someone.

What about consensus? Well this is simply dealt with, truth doesn't care how many people believe it. The frequency that past scientific consensuses were utterly incorrect ought to put down forever the idea that a modern consensus is necessarily right. There is nothing magical or unique about this consensus *this* time. It might be right, but it might also be like the Ptolemaic system of the universe.

Lastly, what about your own effort, your own study? Can you spend enough time on the internet to gain an informed opinion about something like organic food? Unfortunately, to do this without first determining who you will believe is like taking anyone and everyone's word for it. If you spend enough time searching the internet, you will find someone (or many people) on every imaginable angle of a debate. There will be experts and 'experts' and personal opinions and personal experiences. The internet is crawling with frauds like maggots on rotting meat.

So, perhaps the good reader, having made it this far, is suddenly confronted with an uncomfortable question. *How can we actually be certain in our knowledge?* Welcome to the insanity inside the mind of a scientist. Indulge me with a chemistry example. (Remember that I hold a doctorate in chemistry from a well-respected institution… my credentials…)

Let us talk about the atomic structure of an atom. We will begin with the Bohr model. In Figure 11, you can see the nucleus of the atom in the middle (with the + sign that indicates in this case four positive charges) with rings around it that look like planetary orbits. This model makes sense out of the odd observation that atoms only emit light in certain wavelengths. This model says that the electrons may be in any orbit, but never between orbits… electrons may jump between orbits if they absorb/release *exactly* the right amount of energy. (Scientists say that the energy levels of electrons are quantized.)

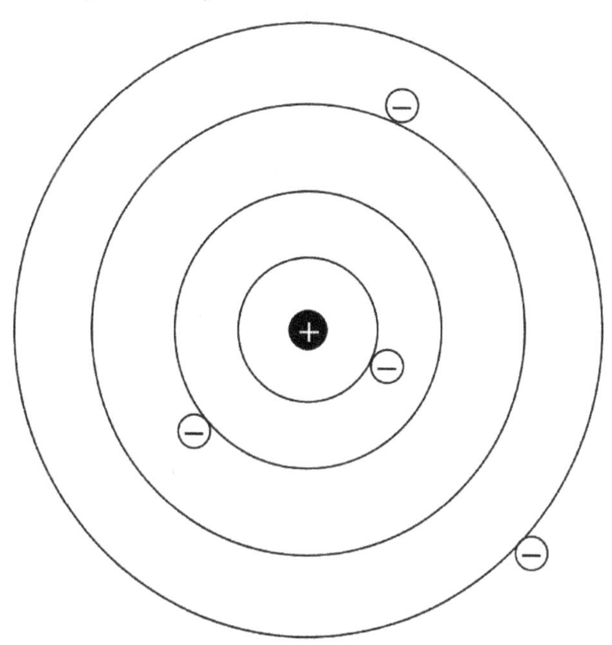

Figure 11: Simple Bohr model of an atom.

This model is simple, clear, and explained the data available at the time quite nicely. Unfortunately for Niels Bohr, his model is almost entirely wrong. The electrons do not orbit in fixed locations, in fact, the electrons are not on one particular place at all. The electrons are in what is known as probability densities of fixed energy levels that can be found by solving the Schrödinger equation. So the lowest energy orbital is called the '1s' and it is a sphere, and the next orbital is the '2s' and it is also a sphere, but a bigger one. The next energy level is called a 2p orbital and it has three orbitals of the same energy and they look a bit like a dumbbell (Figure 12).

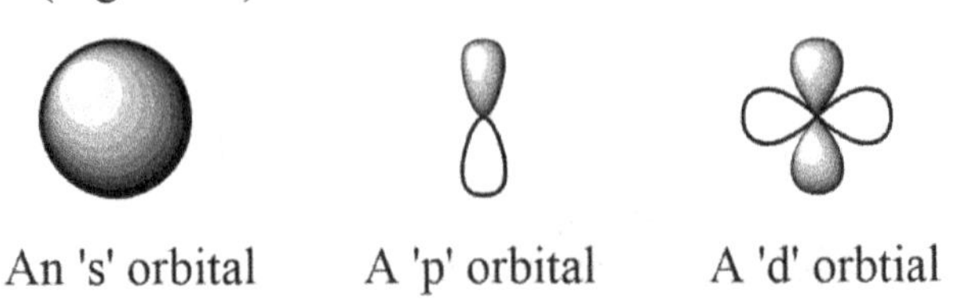

Figure 12: Example s, p, and d orbitals. Note, there are actually three p orbitals and five d orbitals.

The fact that these probability densities still have a quantized (or fixed) energy levels, means that the unique emission spectra of atoms is still explained.

How, do I know these things? Remember, I am a credentialed chemist. I have a bachelors and a doctorate in chemistry. Yet I know these things largely because someone else told me. I also know many of the evidences and scientific experiments done to prove/disprove these theories, but I have not done them myself. I am taking other people at their word. Eventually, while you can know for yourself if the shower water is ice cold or nice and toasty warm, most things you know are either from accepting someone else's authority or taking an authority's premise, and building new work to your own conclusions. This is why it is critical to select carefully those whom you are willing to accept as authorities.

I don't have a good answer for how to tell who is a good authority and who isn't. I do know that much consideration and critical thought ought to go into accepting anyone as an authority, *especially* those people that you have an emotional agreement with. There are a number of markers of untrustworthy authorities. For example, you should be skeptical of those who raise money for things, especially if they develop a track record of raising millions of dollars and yet having cost overruns. Be careful of those who are eager to connect on an emotional level. Watch out for people who use words to mean different things in the same statement, or use common words to mean things not usually meant. And be wary; for the world is full of devious people you think agree with you and yet contains honest people who disagree with you entirely. My general rule is to introduce more doubt about the authority of the people you agree with and don't lose your natural skepticism for those you disagree with. This is a lot of skepticism, and I am not certain that high levels of skepticism are healthy for many people, so you could find a trustworthy skeptic friend, a MacPhee from *That Hideous Strength,* who can let you know of skeptic alarms that go off naturally in his mind. Basically, make a scientist friend or two, and run stuff by them from time to time.

Study work:

1. Make a list of things that you know from your own experience.

2. Make another list of things that you know largely or entirely due to someone else's statements.

3. Take every item in #2 and investigate the authority of the person whose word you have accepted.

4. Select 2-3 of the authorities from #3 and critically evaluate them by the following criteria:
    a. What are their credentials?
    b. Have they run any major projects that overran their budgets by a significant (>15%) margin?
    c. Do they function largely on donations? If so, what percentage of time do they spend on fund raising?
    d. Do they talk about anything that you know from either your own experience or a quality authority? If so, are they right?

Essay 8: √
How Big was the Ark?

How big was the Ark? Could it possibly contain all the animals? Recently some folks have built an Ark replica in Kentucky. There are convenient signs about the dinosaurs being juveniles, so that's how they would fit on the ark. There was much measuring of many cubits. "This is how you are to build it: The ark is to be three hundred cubits long, fifty cubits wide and thirty cubits high. Make a roof for it, leaving below the roof an opening one cubit high all around. Put a door in the side of the ark and make lower, middle and upper decks." (Gen 6:14-16) This sort of display makes a demand on scientists to accept the account as scientifically verifiable.

To step back a moment, the Bible teaches two branches of knowing God, natural and revealed. This division also applies handily to various claims in Scripture. Some events are explicable through observation of the natural order and human nature followed by reasonable deductions. For example, it doesn't take a stretch to reasonably understand the causes of Absalom's betrayal of David, or Saul becoming an evil king. The other kind of event must have its cause revealed, as it is an impossible event. A great example of this is when the sun stood still at Gibeon (Joshua 10:12-14). The first is an event that could be considered scientifically verifiable. We could potentially find artifacts that date from that era, date them with C-14 dating and other methods, and conclude that events unfolded as reviled in Scripture. What science cannot answer is how the sun could stand still over Gibeon. A miracle is the direct and unmediated action of the Creator on His creation.[14] These events are, by the very nature of

[14] I feel strongly that this formulation is from someone, somewhere. Perhaps it is a well memorized definition given to me during the catechesis of my youth, but I cannot find this direct quote anywhere… the closest I can find is F. R. Tennant's *Miracle and its Philosophical Presuppositions: Three Lectures. Delivered in the University of London 1924*, pg 48. But I am nearly entirely certain I have not read this work before. It is now on my list…

science and human reason, scientifically inexplicable.

Let us consider a moment the Flood story.

> So God said to Noah, "I am going to put an end to all people, for the earth is filled with violence because of them. I am surely going to destroy both them and the earth. So make yourself an ark of cypress wood; make rooms in it and coat it with pitch inside and out. This is how you are to build it: The ark is to be three hundred cubits long, fifty cubits wide and thirty cubits high. Make a roof for it, leaving below the roof an opening one cubit high all around. Put a door in the side of the ark and make lower, middle and upper decks. I am going to bring floodwaters on the earth to destroy all life under the heavens, every creature that has the breath of life in it. Everything on earth will perish. But I will establish my covenant with you, and you will enter the ark—you and your sons and your wife and your sons' wives with you. You are to bring into the ark two of all living creatures, male and female, to keep them alive with you. Two of every kind of bird, of every kind of animal and of every kind of creature that moves along the ground will come to you to be kept alive. You are to take every kind of food that is to be eaten and store it away as food for you and for them."
>
> Genesis 6:13-21

Then the rains came down and everything with breath, not on the Ark, died. Could God have saved Noah's family and the selected animals some other way? Say, making a sky envelope and lifting everyone up and out until the flood was past? If you believe that God is omnipotent, then you would have to agree that, yes, God could have saved Noah in this fashion. However, God chose to have Noah build a large ship. It seems that, since God could have saved Noah and the animals in any way he wished, the primary reason to use an ark was not to save Noah and his family. That they were saved was secondary to the ark-building. The Ark must have served another purpose, one that would not happen if, at the start of the rain, God took Noah and a lot of animals into occlusion to release upon the cleansed earth.

In 1 Peter 20-21 we are told that "God's patience waited in the days of Noah, while the ark was being prepared, in which a few, that is, eight persons, were brought safely through water. [21] Baptism, which corresponds to this, now saves you". This indicates that the Ark is a type of the Church and the Flood is a type of Baptism. Noah was undoubtedly a prophet to his time. He also received very detailed instruction from God on how to build the Ark. In his *Commentary on Genesis* Luther notes that the instructions to build the ark are carefully made, as with the instructors for the tabernacle.[15] Luther's point is that to follow the detailed instructions involves greater faith and lessens the doubt that the work is ordered by God. It is also consistent with the ark being a type of the Church.

It is not a significant intuitive leap to conclude that Noah built the ark primarily as a witness and prophecy to the wicked and damned generation he lived among, as well as a prophecy of (as detailed in 1 Peter) the Holy Church. Noah was essentially a prophet to his generation of the righteous condemnation of God, but unlike Ninevah's repentance upon the prophecy of Jonah, the people of Noah's day did not repent. One cannot believe that, if someone of Noah's time had believed Noah, repented in his equivalent of sackcloth and ashes and fled, trusting God, to the Ark, that God would not have rescued him.

So since the Ark was not primarily an instrument to rescue a bunch of animals, but a witness to the unbelievers, the Ark is quite possibly as miraculous as the Flood itself. While one may build a replica ark, and do a bunch of math projecting the number of species (kinds) of animals and the amount of food required and with some clever stacking and packing make a claim that they would all have fit upon the ark, this is unnecessary. Perhaps it is even counterproductive. As soon as you begin explaining a miracle with math, then you require one of two things. You either permit the possibility of being disproven by math (live by algebra: die by algebra) or you require intellectual dishonesty from believing scientists. You cannot insist that an event is

[15] Luther, Martin *Commentary on Genesis: Luther on Sin and the Flood.* Page 249.

scientifically verifiable while simultaneously discarding any scientific evidence. Well, you *can* but it isn't recommended.

(This principal applies to the floodwaters themselves as well. While it is feasible to at least partially explain where the waters came from and where they went, to do so requires that you accept objections to your narrative based on observations.)

Another issue is that, while the size of the Ark is relatively fixed, the number of animals, the amount of food they need, and their respective sizes, at the time of Noah, are all wild estimates. As someone with a lot of experience making scientific estimates, I can testify that estimates are… largely inaccurate. (They are useful so that you predict the thermodynamics of your reaction, but you can be off by quite a lot… and still avoid blowing up the place.) In a system such as the ark, there is the added issue that one estimate influences the other estimates. This means that if the first one is wrong (how many species there are) the next one will be very, very wrong (how much food is needed).[16]

Perhaps the most fundamental issue is that the Bible has one purpose, to teach mankind about God and specifically about Jesus Christ and Him Crucified. To use the Bible as a scientific field guide is as much a subversion of its purpose as using it for financial advice or a 'Scriptural weight-loss' program. There is nothing indicating in Genesis that the Ark was 'bigger on the inside', but it is wrong to convert a miracle into an event which is open to scientific observation. The Ark may have contained more creatures than it ought just as the widow's oil jar contained more oil than it could.

Study Work: This topic is closely related to the next topic, therefore the study work is presented after the next essay (and it is longer than otherwise).

---

[16] Also, please leave out any speculation of 'juvenile dinosaurs' or 'dinosaur eggs'… the Bible clearly states that mating pairs were taken into the ark. (Gen 7:2)

Essay 9: The Age of the Earth

Before we get into this topic, I feel the need to insert a disclaimer. This is important as you read this essay. I believe in 6 day creation as taught by the Lutheran Church – Missouri Synod. I have experienced on numerous occasions that this topic draws much emotional response, and as such I felt it prudent to begin with that statement.

The age of the earth and the theory of evolution are often conflated in the minds of evolutionists and creationists alike. While evolution does indeed depend on a very old earth (perhaps older even than the 4.5 billion years claimed), the old earth does not depend in any way on evolution. From this set of logical facts, we can clearly and safely separate the issues into two distinct debates. While evolution is discussed in another essay, here only the age of the earth and the age of the universe will be discussed.

The Bible clearly does not leave much time after the Fall in Genesis to the time of Christ, which is historically verifiable of being about 2000 years ago. The exact date calculated by Bishop Ussher of creation occurring at 4004 B.C. is not supportable by Scripture as Cainan, mentioned in the New Testament genealogy, is not mentioned in the Genesis genealogy upon which Ussher based his calculations.[17] However, a very young earth is the most consistent interpretation of the timeline presented in Scripture.

On the other hand, there are a number of observations made of the universe which all seem to clearly, and consistently indicate a universe of many billions of years old, and an earth of a few billion years old. The evidences given include: 1) the expansion rate of the universe 2) starlight 3) the existence of numerous isotopes of nuclei. Only the last item has even a tangential relationship to my expertise in Chemistry.

I believe that the primary requirement in discussions, especially

---

[17] Klotz, John W, *Modern Science and the Christian Life*, pg 116.
Jurchen, John, Concordia journal, Volume 43, number 3, pg 70.

on this topic, is academic humility. Both theologians and scientists ought to approach this issue from their respective domains with caution and without either pronouncement of hearsay or of ignorance. (You will notice in the sections on evolution that I make no such admonitions, as evolution is far less scientifically viable and has far less concrete observation in support than do these ancient earth measurements.)

Both kinds of humility are difficult and have slightly different appearances. In the first place, a new idea in science has a decent chance of being the best idea yet. From a heliocentric solar system to general relativity to the quantum mechanical model of the atom, it is clear that, while new ideas can be crackpot theories and garbage, they also are often really quite accurate and beneficial. However, in theology, the new idea is essentially always heresy. (Incidentally, this might be a major source of friction between the two communities.) However, not everything that churchmen think is actual doctrine, and not every idea espoused by theologians is eternal truth. A long time ago, the church was confronted with a scientific theory, Ptolomy's planetary model, and decided that it was right. Then the church proceeded to read their theology with that model in mind, until it came to pass that the church conflated believing that the earth was at the center of the solar system with believing the Word of God itself. Passages of Scripture were used to defend the geocentric model which the churchmen assumed were being directly denied in an act of denying Scripture when the heliocentric model was proposed. Nowhere in Scripture does it say 'the earth is at the center of the solar system' and yet there were theologians who believed it was utter hearsay to claim otherwise.

(Before we continue, please remember the disclaimer at the beginning – 6 day creation… LC-MS teaching… etc… thanks)

The mistake here, is that one particular theory about the natural world (making it a theory that science is expected to have a valid opinion on) is used to interpret odd passages of Scripture, such as the sun standing still over Gibeon in Joshua 10. This fits nicely into Ptolemaic geocentrism, but is odd and difficult to explain with heliocentrism. Then, over time, the theologians mistook their fitting of Scripture to scientific thought as theological truth in itself, and from that we have the accusations of heresy leveled at Copernicus.

Think about it this way, we now say that (since the earth clearly and undeniably orbits the sun) the account of Joshua 10, where it clearly says that the 'The sun stopped in the midst of heaven and did not hurry to set for about a whole day' (Joshua 10:13b) that the Bible is using 'phenomenological' language.[18] What this means is that God, in mercy and to make Himself understandable to man, will allow events to be described, not as they were accomplished by God, but as they were observed by man. Since the Sun appears to rise and set around the earth, the eye-witnesses in Gibeon that day would say that the sun stood still, even though the sun didn't stand still. Imagine, for a moment, trying this explanation on a 16th century theologian, and I would greatly expect that you would be castigated in a way that would make modern condemnations appear to be sweet and delightful conversation.

So that is why I recommend theologians approach this issue with humility. We believe that the best reading of Scripture fairly clearly indicates that the earth is young, say no more than 10,000 years, probably not much more than 6000 years. However, the age of the earth is itself a phenomenon that science can actually investigate, and so the theologian ought to remember that just because there is not currently a scientific theory which is harmonizable with Scripture, there may be one in the future, and it may not be exactly what they think now. It is entirely possible that, as time is a created entity itself, the 'time' of creation is not something understandable by human minds.

As for the scientists, the case for humility is easier. The history of science is a long litany of brilliant men being terribly wrong. The impressive part about science is that, in the Christian West, there has been just enough self-criticism to recognize that the brilliant are often wrong and eventually we discard ideas like the Bohr model of the atom and the geocentric solar system. Scientists today are rather set in their ways. Despite the hype, scientists are some of the most close-minded people I've ever met. I have seen full on arguments between high level scientists over the energy

---

[18] Harstad, Adolph, *Joshua*, Concordia Commentary, Pg 423.

level order in a molecular orbital diagram. You probably didn't understand that, but that's my point, it is not something an open-minded person would get angry about. But, every scientist in theory at least would accept the possibility that (after much ridicule and argument and huge amounts of corroborating evidence) it would be possible for science to turn around in 50 years and say 'sorry guys, we screwed up… the universe is 20,000 years old… max'. If any scientist argues against even the possibility for a total rewrite of physics… maybe check and see if your scientist is an economist who makes computer models for a living.

So what does that have to do with the age of the earth? (Remember before you continue, the disclaimer at the beginning, about me believing in a young earth – 6 day creation…) I think there are a few critical conclusions. First, I will exercise a level of academic humility myself and say this: I do not know how these disparate views hold together but I know that God does not lie and He does not play tricks. So I reject completely the idea that God made the universe to look old for fun. I believe that God made the universe whole and complete, e.g. He made an 'adult' universe in five days and then, on the sixth day he made an adult man, out of dirt. The fact that the science does not seem to coincide with the biblical teaching doesn't really bother me, since science is often wrong, science cannot investigate the miraculous, and while our understanding of Scripture is flawed since we are flawed, Scripture itself is True.

I do not think arguing or finger pointing over this issue is useful or helpful. There are many Christians, largely in the sciences, who struggle with issues like this, and they deserve the respect and support of their clergy and their brothers and sisters in Christ as every sinner struggles with his own doubts. As C. S. Lewis would not harshly criticize a gambler since he had no temptation to gamble (I think found in God in the Dock…) so those who find no intellectual struggle in the age of the earth debate ought to not be overly critical, but encourage their brother in his faith in Christ and belief in the Scriptures. (How one should deal with arrogant academics… hopefully of which I am not a part… is perhaps another issue.)

Study work for Essay 8 and 9:

1. Consider the following topics:
    a. The Virgin Birth
    b. The Resurrection of Jesus
    c. That Jesus had a physical body
    d. The genuine existence a first man named Adam
    e. The mechanism by which Jesus turned water into wine
    f. How God parted the Red Sea
    g. The Trinity
    h. Faith without works is dead
    i. Did Lazarus actually die?
    j. Jonah in the 'big fish'
    k. the nature of the Christmas star.

Sort them in a column down the left of a piece of paper by importance to the faith. Then put boxes with the following three categories: Critical, Important, and Irrelevant, around the appropriate sections of your list. Imagine someone does not agree with you about each one, and indicate in a second column if that disagreement ensures their lack of salvation. In other words, would you be willing to call (or strongly imply…) someone a heretic for disagreement? Where is the dividing line? For each point look through Scripture to see if you have sorted correctly and cite as possible from Scripture. Where not feasible put down a logical case for your judgment.

2. Does the Bible always clearly indicate when an event is miraculous and when it is a physically explainable?

3. What is the role of humility is having discussions, especially heated ones?

Interlude

The Disappointed Alchemist: Dylan J. Thompson:

I becams't an inorganic chemist
Haphazardly, by promises seduc'd
(Spoken softly, by Sybil who liest)
Of brilliant colours to essence reduc'd.
I had thought to create sanguine, azure
Sable, verdant compounds; myster'ous, whole.
As hoary Alchemist of old, hunch'd o'er
Sought full transmutation of leaden soul,
I had thought to find mystery and subtlety.
Recalcitrance, instead, and thoughtless perfidy
In great supply I find. Now, my soul sees.
Lo! Mankind hath lost all but rational
Deceit, which, as Circe kept Ulysses
Does captivate and mesmerize the soul.

Essay 10:
Science 'News' and the Philosopher's Stone

Long, long ago, when chemists were still Alchemists, and real men wore beards and magnificent mustaches (looking *right at you* Tycho Brahe), brilliant and methodical men wasted an inordinate amount of time searching for the philosopher's stone. This, for anyone without any fantasy knowledge, is some material (not necessarily a stone) that would transmute lead or other base metals into gold. Often, this was also used as a metaphor for the philosopher/alchemist who sought some way to transmute their base instincts into noble characteristics.

Figure 13: Tycho Brahe's mustaches.[19]

The unattainable dream of the philosopher's stone actually

[19] Wikipedia Commons.

drove much invention, and yet it also caused much wasted time. I don't mind that really, since alchemists were funded like any artist at the time, by people with money who wanted to look cool. Today's alchemists, however, use taxes to fund their dreams, and today's newsmen both do not understand scientific results *at all* and they have certain policy/belief agendas with which they filter every science news story. To illustrate this, we can look at one branch of research which is very common, and seems to be a modern philosopher's stone of sorts: hydrogen power.

The appeal is certainly very large. Combustion, the simplest way to eke power out of nature is the rapid oxidation (with oxygen no less) of a chemical. For instance, the combustion of octane is written chemically like in Figure 14: (where octane is the zig-zag, all carbons are points on the lines and the hydrogens are implied… chemists are a bit weird.)

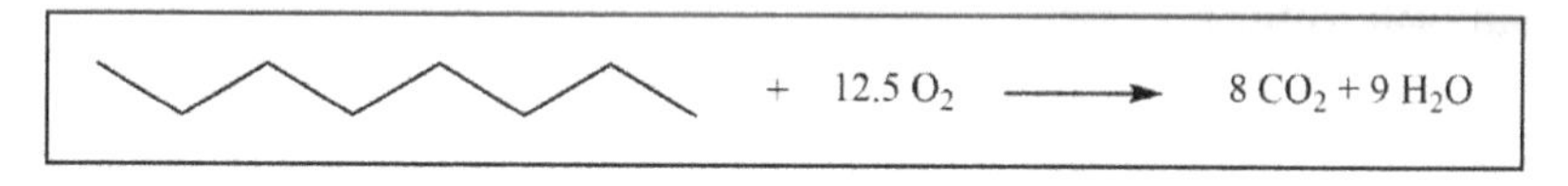

Figure 14: combustion of octane

This produces in an ideal situation, lots of carbon dioxide. (How bad this actually is, is a different discussion, also tainted by the policy beliefs of the scientists and reporters involved.) To make it worse, the combustion of gasoline is never this tidy. There are a massive number of different hydrocarbons in gasoline, some of which (a very few these days) contain sulfur, nitrogen etc., giving as product (or exhaust) oxides of nitrogen, sulfur, and carbon monoxide. In comparison, hydrogen combustion is surprisingly tidy. Ideally, it yields no products other than water (Figure 15).

$$2H_2 + O_2 \longrightarrow 2H_2O$$

Figure 15: combustion of hydrogen

That means power with (ideally) *no pollution whatsoever*! Which is, admittedly, a very attractive idea. Now, while technically possible, and definitely appealing, every chemist will readily admit that there are serious drawbacks. However, this book is not specifically for chemists. Hopefully non-chemists will read it, many of whom have probably at some point read glowing articles in Popular Mechanics or Popular Science, or some article on the economy predicting the coming dominance of the hydrogen economy. For those people, I want to lay out one of the theoretical difficulties with the whole idea. There are a number of problems which I omit because this is an illustration not an article on the difficulties of the hydrogen economy.

Problem: Where do we get hydrogen? Well, right *now* we get hydrogen by putting a lot of energy (probably from coal) into… methane… fossil fuels. This means that currently, to use hydrogen as a fuel we would have to take a pre-existing fuel, use energy to lower the energy density of the fuel, and then use the new worse fuel. Not something anyone other than a bureaucrat would do. (This is like using Ethanol in cars… I think I should avoid that and stay on topic.)

Theoretical solution: This is the solution that is the most like hunting for the philosopher's stone: water splitting. The idea is this, use solar energy (it has to be not fossil fuel based to make any sense, so solar or nuclear energy is about it.) to drive a catalytic reaction that splits water into hydrogen and oxygen. The technical difficulties of water splitting are huge and not really comprehensible without a few thousand words of explanation. However, in a nutshell, there are two distinct steps to the process, the hydrogen side and the oxygen side. The hydrogens in water are formally +1, and so two hydrogens, in order to form an H-H bond (hydrogen gas) need to gain an electron each. The oxygen in water is formally 2^- and to form O=O (oxygen gas) the need to lose two electrons each. Right off, you notice that the number of electrons match, that's the easy part. The problem is that it is immensely difficult to get oxygen to give up its electrons once it has them. (In chemistry talk, we can say either that the electron affinity of Oxygen is very large, or that the thermodynamics of O^{2-} formation are very favorable.) What this means is that currently, in order to oxidize the oxygen you have to use something that wants the electrons even more than the oxygen does (and that is most definitely *not* the hydrogen). So the current situation is that experiments are essentially a proof of concept where either the H^+ is reduced to H_2 gas or the oxygen is oxidized to O_2 gas. These methods use the massively expensive, typically non-reusable catalysts such as ruthenium, rhodium or iridium based catalysts and large stoichiometric amounts of redox reagents to either donate or take the electrons (depending on which side is being studied).

Looking at the proposed 'solutions' to me, the whole water splitting for hydrogen fuels looks like a big impossible circle (Figure 16). While this is not, technically, an error in the line of perpetual motion machines, (there is energy input from the Sun, which powers the cycle), the impracticality of all steps is largely not discussed.

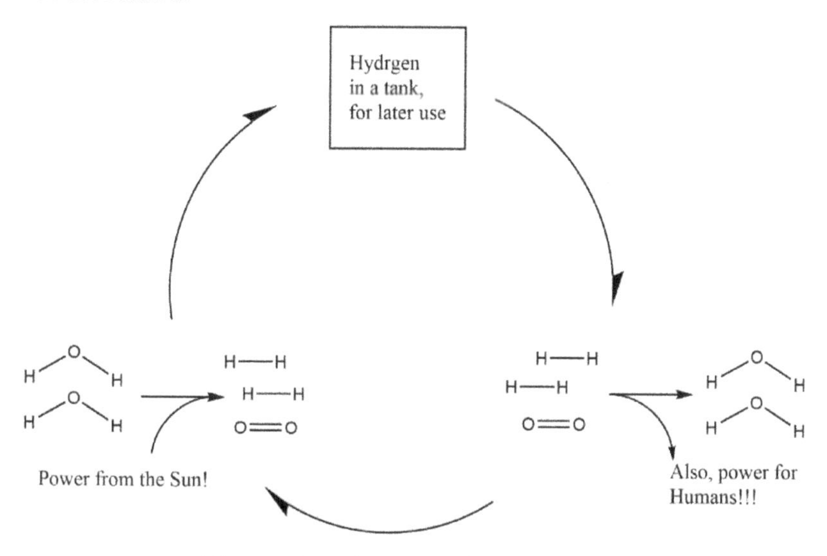

Figure 16: Hydrogen Fuel Energy Cycle

Now I fully and unequivocally believe that scientists should attempt the impossible. Many times what seems impossible becomes possible because stubborn hard-headed scientists keep trying to do them. That is perhaps one of the most admirable things about science, persistence in the face of overwhelming odds.

What you need to watch out for is when people, scientists included, behave as if their experiments earth-shatteringly important. When a news article proclaims the upcoming age of ‘clean power’ or the discovery of the earliest life on earth or the ‘very likely to find evidence of life on mars’ readers ought to show a healthy skepticism of the claims of both the scientists (who have ideological and financial reasons to inflate the importance of their work) and the science ‘news’ that is almost exclusively misunderstood in favor of ideas like: ‘oil is bad’, ‘creationists are dumb’, ‘evolution is a proven fact’, ‘Global warming is real and don’t you dare ask a follow-up question’. Remember, there is really no such thing as a non-partisan or unbiased observer. But perhaps the worst possible source of real information is ‘science news’ where the ‘reporter’ doesn’t know what is going on, and the scientist likes the fabulist spin or is too busy to comment. Its best to remember that though expertise means something (and not everything), what comes in the news is filtered through the mind of a non-expert, and my experience is one where the news article makes claims that are entirely absent from the scholarly work they are written about.

Study Work

1. Do a search for news articles about the age of life on earth or about the connection of autism with vaccines. Refine your initial findings to those that mention or cite an actual academic paper. Usually, with or without a library subscription to articles, you will be able to access the abstracts, which are like executive summaries of the work. Compare the claims made in the news articles to those made by the scientists.

2. Scientists sometimes overstate the importance of their research in the introductions to articles. Search for a scholarly article published in a chemistry focused paper that deals with 'water splitting'. Read the introduction(s) carefully, and count the overstatements and compare the number of times that difficulties are discussed. If you have access to ACS journals, use those, otherwise there are a number of decent journals with some open access publishing, and you should be able to find something. With minimal poking around, I found a *PNAS* (Proceedings of the National Academy of Science) article about this very topic that is publically available as of this writing.

Essay 11: √
The Great Extrapolation

Science is so frequently abused it is no wonder that people are rapidly losing their trust of anything with the label science.[20] I suppose this is the just end for those who used the science label and the trappings of science, graphs, charts, R^2 values, big words, to trick people who rightfully trusted science. Science, real science, is a method to discover the truth about the world. In this system it is acceptable to make extrapolations and theories that are difficult or even impossible to prove. This creative side of science gives scientists and people ideas to test, and if real science is being done, discard if they are found in contradiction of observation. As a result, any theory that *cannot* be tested is in essence *not* science.

One of the most common features to the 'sale' of pseudo-science and loony conspiracy theories dressed up as science is the graph; specifically the false extrapolation graph. Here, I think, is a perfect example of what I mean. When I was small, my mother and father kept track of my height at different ages. When I plot this data against my age in years, the resulting graph is below.

Figure 17: Growth Chart

[20] This essay is based on two posts on my blog dating July 16, 2012 and April 18th 2013.

As you can clearly see, my growth between age 11 and age 14 was really nicely correlated to my age. The graph tells us that I grew about 4.3 in/year for two and a half years, at a steady rate. The R^2 value of 0.99 is a good statistical correlation, so lets' extrapolate.

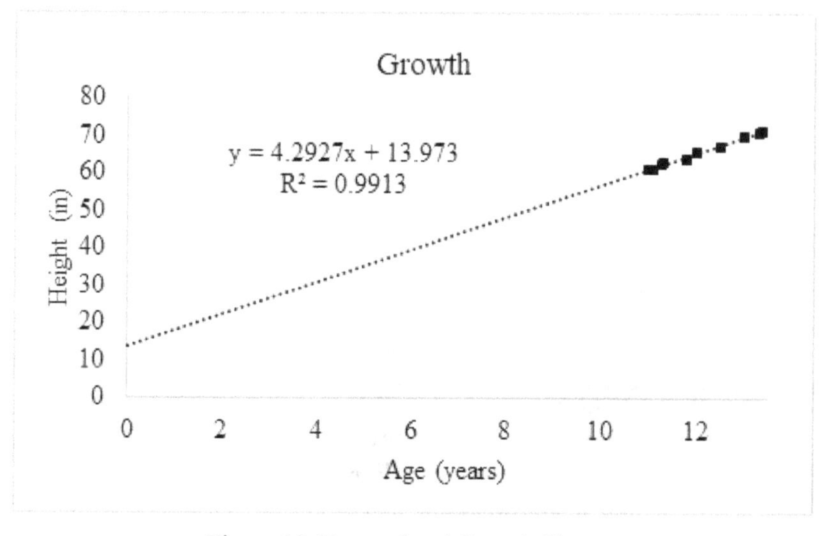

Figure 18: Extrapolated Growth Chart

OK, so for anyone with high school math, the 13.973 inches at age 0 wasn't a surprise. (If x = 0 than y = 13.973 from the equation for the line.) Now that is absurd itself, but let's extrapolate into the future the same amount for really obvious absurdity.

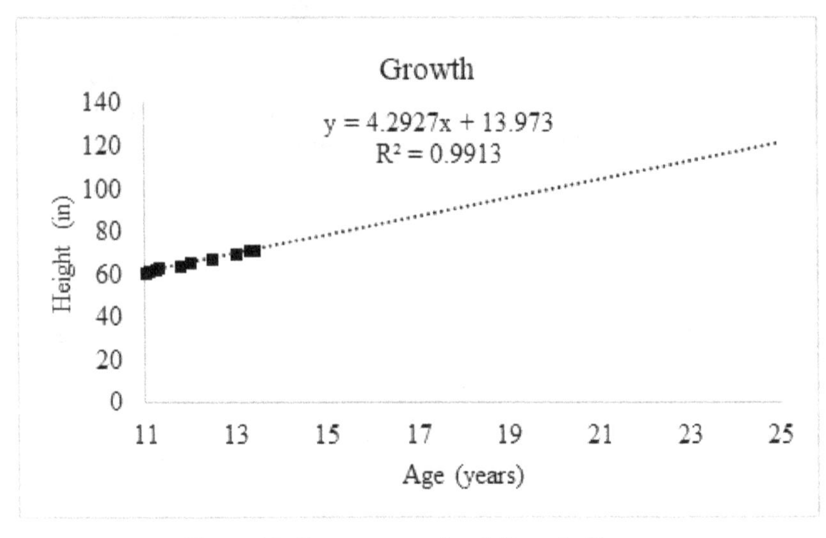

Figure 19: Super extrapolated Growth Chart

That puts me at around 120 inches at 25 years old! (You don't want to know how tall the graph says I'll be next year…) Alright, so this example is obviously ludicrous. These examples seem crazy since they are connected to something that everyone has tangible experience with. However, the same error can be made, both intentionally and unintentionally, in topics that are much more intangible. Before we get into an example of that, let's consider the correct way to use this data.

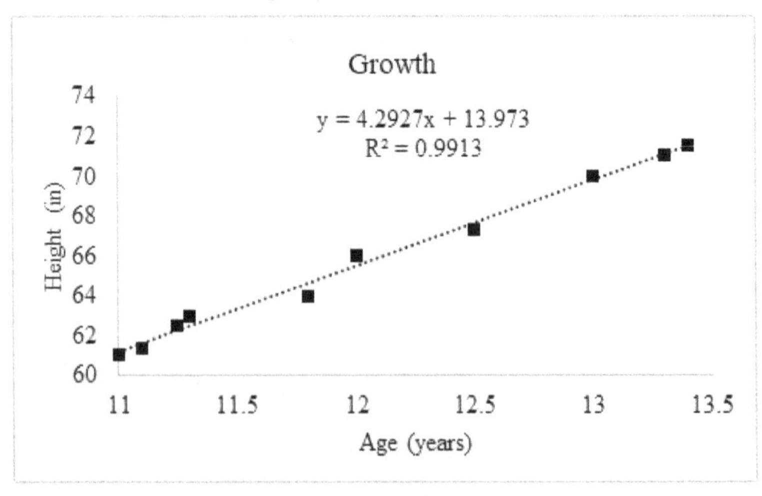

Figure 20: Proper calibration plot

There is one valid way to use the information. If we calculate some unmeasured age, say when I was 12.7 years old. And use the linear fit at the top, we can calculate that I was 68.45 inches. This calculated height is reasonably accurate to my actual height at that age. The basic rule is that the correlation found from a linear relationship between two variables (height and age in this case) is **only** valid inside the range of measurements. Further, it is also most accurate towards the center of the measurements.[21] In this case that means that we can use the relationship from 11 years to a little more than 13 years. Outside the observed data, the relationship is highly likely to break down. Science is messy.

It is not entirely wrong to use an observed trend to extrapolate future events or past conditions. However, these calculations are far less meaningful than the weight usually given to them. Science exerts itself to make predictions in order to test theories. If a theory allows for predictions to be accurate, that is good evidence for the theory. In the case that the predictions are invalid, then the theory is either wrong outright or it is seriously flawed and needs to be reconsidered.

Outside this application, the use of mathematical models and data fitting ought to be treated with skeptical eyes. Unfortunately, the misuse of mathematical fitting or modeling is a common feature of politicized scientific disciplines.

An example I found of this type of 'graph abuse' is from an MIT Technology Review post about a correlation between organism complexity and evolutionary timing. Now, the article[22] begins with a brief discussion of Moore's Law. This states that the number of transistors on a computer chip will double about every two years, and so eventually the transistors will become too small for lithography to work. This will result is a computational power 'wall'. Well, if you plot transistors on a chip versus time, and look

---

[21] Officially the centroid, or weighted average. If most of the data are at one end of the linear fit, then the most accurate location will not be the center, but towards the end with more data. You can imagine a balance beam, and each datum 'weighs' the same, and the most accurate part of the linear fit will be at the center of gravity.

[22] https://www.technologyreview.com/s/513781/moores-law-and-the-origin-of-life/

at where there should be '0' on a chip, this works pretty well (mostly because we know when the first transistors were built. The '0' point is in our measured data). The article also gives the example of scientific publication quantity and calculating back to right around Newton's time as 'start' for scientific publications. This also is fairly reasonable since the publication quantity versus time is actually all data that we have. Both of these example are like the 11-13.5 year old height plot in figure 20. There really isn't much if any extrapolation involved. However, after those two reasonable examples, the article brings up a rather unreasonable example (Figure 21).

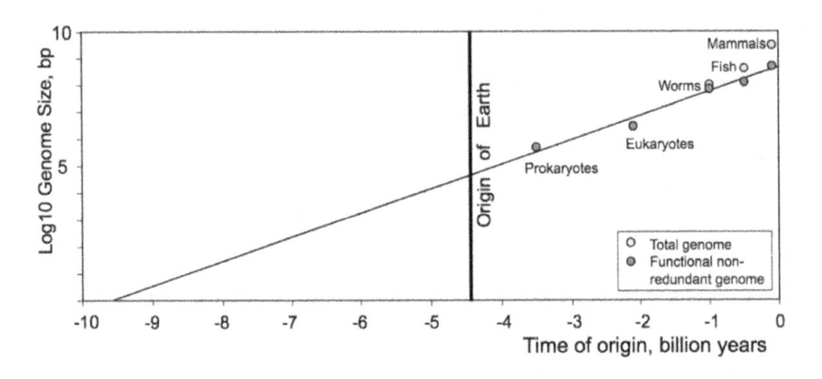

Figure 21:[23,24] Extrapolation chart of the complexity of life.

Here is the problem: the article is about extrapolating back *life* to calculate a 'start' time for life. There are several issues here, and the abuse of the graph takes a couple of forms. The first abusive part of this article is that it conflates measured and verifiable data with unmeasurable and, by definition *unverifiable*, data as if they have the same value. Measured data is like my height at 12 years old or the average number of transistors were on a chip in 2000.

Unmeasured (and unmeasurable) data is something like when prokaryotes first appeared. There are layers upon layers of reasoning for why biologists date Prokaryotes before Eukaryotes, but I have a sneaking suspicion that one of the reasons is that

[23] https://www.technologyreview.com/s/513781/moores-law-and-the-origin-of-life/

[24] Sharov, Alexei A., Gordon, Richard, arXiv:1304.3381 [physics.gen-ph], 2013.

Prokaryotes are simpler than Eukaryotes and therefore 'must' have come first.

It should be redundant that these complexities fit a log scale if, they were dated by a log scale in the first place. Basically, if you were to decide that the appearance date and genome size were related by a log scale, and you then dated by that assumption, the data you produce is simply a graphical representation of your initial assumption. Since there was no one around making observations two billion years ago, the time that eukaryotes first emerged is dated at least partially on our original assumption.

The other issue is that even if we had incontrovertible data to prove when different organisms evolved you still cannot scientifically extrapolate beyond your data points and obtain anything useful. This extrapolation back to 'beginning of life' assumes that everything works the same for *billions* of unobservable years. There is one way that using such a plot would be scientifically correct. If evolutionary dates were measurable (but difficult to measure) you could use it to estimate a new species' 'evolution date' which then might be verified if needed. This example ties back into what we talked about in Essay 10: a science oversell. 'We can calculate when life first evolved!' is very attention grabbing while the reality is that nothing was calculated and nothing was found.

(Of course, if you, dear reader, were paying attention you noticed. If this data were true, we would have to have a mechanism for life to begin billions of years before the earth is supposed to exist.)

This is just one example of some idea essentially wearing a science skin-suit and strutting around pretending to be science. It is important for Christians to be able to tell that apart from actual science. If we condemn the sciences, we are abandoning a powerful tool for the truth, one that Christians invented. Also, we leave the whole population defenseless against horrible ideas that dress themselves up as if they were science. It makes our own Christian brothers susceptible to fraud and scams, and leads our children astray. Science, real science, is a hunt after truth, and is an antidote to this insanity.

Study Work

1. There are many variations of ideas wearing the 'science skin-suit' that are not science at all. These are generally found where the language or mathematics of science are used, but the evidence is thin on the ground. Think of at least three popular things that use the vocabulary and syntax of science, yet are not, in fact, scientific. (Hint: if it advertises 'studies that show', it is suspicious.)

2. While discussing the origin of life / Moore's Law example, I describe what amounts to very fancy circular reasoning. Sketch a figure to illustrate the fallacious circular argument I describe.

Final Interlude

The Gods of the Copybook Headings
(Rudyard Kipling)

On the first Feminian Sandstones we were promised the Fuller Life
(Which started by loving our neighbour and ended by loving his wife)
Till our women had no more children and the men lost reason and faith,
And the Gods of the Copybook Headings said: *"The Wages of Sin is Death."*

As it will be in the future, it was at the birth of Man
There are only four things certain since Social Progress began.
That the Dog returns to his Vomit and the Sow returns to her Mire,
And the burnt Fool's bandaged finger goes wabbling back to the Fire;

And that after this is accomplished, and the brave new world begins
When all men are paid for existing and no man must pay for his sins,
As surely as Water will wet us, as surely as Fire will burn,
The Gods of the Copybook Headings with terror and slaughter return!

To Them That Mourn

(G. K. Chesterton)
(W.E.G., May 1898)

Lift up your heads: in life, in death,
God knoweth his head was high.
Quit we the coward's broken breath
Who watched a strong man die.

If we must say, 'No more his peer
Cometh; the flag is furled.'
Stand not too near him, lest he hear
That slander on the world.

The good green earth he loved and trod
Is still, with many a scar,
Writ in the chronicles of God,
A giant-bearing star.

He fell: but Britain's banner swings
Above his sunken crown.
Black death shall have his toll of kings
Before that cross goes down.

Once more shall move with mighty things
His house of ancient tale,
Where kings whose hands were kissed of kings
Went in: and came out pale.

O young ones of a darker day,
In art's wan colours clad,
Whose very love and hate are grey—
Whose very sin is sad.

Pass on: one agony long-drawn
Was merrier than your mirth,
When hand-in-hand came death and dawn,
And spring was on the earth.

Essay 12:
The Enemy and His Heavy Cavalry

Thinking back to the battle-lines analogy, it is critical for Christians to assess the situation around themselves, and accurately identify not just current zones of conflict, but rather where the Enemy is maneuvering. And yes, the capitalization is intentional. Christians often forget, in practice if not in theory, that the Devil is a person, and has legions of mighty followers. The Devil and his demonic minions make plans. They set out to devour us. I remember an excellent children's sermon where the pastor had a target on his back. That is the reality and we often behave as if people going around and doing bad things, or advocating for wicked ideas is simply just Human nature. (Obviously human nature is sinful form birth, that's not the point here.)

Think for example, about demon possession. The Modern Christian behaves as if that was an entirely 1st century phenomenon. Reading in *Christian Dogmatics,* the description of demon possession is very much a present tense issue. For example, talking about 'bodily obsession' Mueller states that it 'occurs when the devil immediately and locally inhabits and governs the body, controlling it according to his will, Mark 5,1-19; Luke 8,26-39'.[25] I suspect that many readers will agree with this, however, the practical application of this recognition of demonic activity will sound very *very* 'unsciency'. And, since we live in a world where one of the guiding religions is essentially a cult of scientism it is difficult to not adopt some of the prejudices of the world around us.

When evaluating emerging threats, it appears that the Enemy's heavy cavalry is already in fully career charging our weak flank, they have already overrun our outer defense, and threaten to fold up the whole line. While I am sure that there are other upcoming issues, one in particular that I think is entirely underestimated by the Christian world is the one from the LGBT ideology. I get the

---

[25] Mueller, John Theodore Th.D., *Christian Dogmatics* 208.

impression that Christians always show up a few decades late to the battle-lines. For example: evolution was ingrained entirely in many churches and all of the secular world without much of a second thought. Now, there are Christians who have at least stemmed the onslaught. It is unlikely that Concordia University Wisconsin will get shut down by either the government or angry mobs because we do not ascribe to evolution. Our stalemate is that society thinks we are backwards science haters, but we largely get left alone to teach what we believe. (This is despite the fact that evolution as a theory has more holes than my colander.) The problem now is that it seems that the orthodox (shall we call them hermeneutically faithful?) Christians seem quite happy to develop arguments about the interior volume of the Ark, and where all the water for the flood could have come from, and are quite unaware of the onslaught of 4th wave feminism and the transgender ideology that is a figurative heavy cavalry charge to the flank of unprepared foot soldiers.

Before we move on into some introductory discussion of feminism and transgenderism, I'd like to give a brief note to the reader. I hope you bear with me, and perhaps read a few of the works in the bibliography that I cite here. Remember, just because something makes you angry, doesn't mean it isn't true. Also, as you read, you may find things out that you have never heard or read before. Then, I hope, rather than take my word for it, or dismiss it out of hand, you hunt it down and see for yourself. What is the core principal of feminism? Is it equal rights under the law? (Which may be your working definition.) But more important that that question, is this one. 'Whose definition is the one that *matters*?' Notice that I did not say 'who is right?' Our secular society is composed of Nietzschean power-seekers. What is right is irrelevant to them, what is useful and what brings power is what matters. It is important to investigate the actual goals and power structures of feminism since it has been so absorbed by Christians for the last fifty years that many are entirely oblivious to the amount of feminism that they believe that is contrary to scripture. There is a modern world syncretism between Christianity and the feminist movement that makes this discussion difficult.

Keeping in mind the difficulties, I hope to convince my readers to investigate this idea: that the definition of feminism in your head is intentionally there, by incessant propaganda and persuasion, for its usefulness to those whose working definition of feminism is something entirely different. And these are the people whose goals and desires are furthered and assisted by the mass number of people who deem themselves 'feminists'.

So, whose opinion on what feminism is matters? The opinion of the people teaching feminism to the next generation of girls. If you go to tumblr, and search 'feminism witch' I believe that you will be more shocked and more sickened than if you had searched for pornography. You will see comments such as 'Lesbianism is a gift from the goddesses Themselves' and 'Gay witches for abortion' (by the way, 'feminism lesbian' produces a prodigious number of results). Sure, it can be argued that these people are not representative of feminism. However, I can share with you, thanks to the investigative prowess of Robert Stacy McCain in 'Sex Trouble'[26] a variety of quotes directly from leading feminists, both from the 60's and from gender and women's studies departments on college campuses. Here are a few exhibits.

Exhibit A: Kate Millett.

From her Wikipedia bio she 'was a leading figure in the women's movement, or second-wave feminism, of the 1960s and 1970s. In 1966, Millett became a committee member of National Organization for Women and subsequently joined the New York Radical Women, Radicalesbians, and Downtown Radical Women organizations.'

In an article entitled **'MARXIST FEMINISM'S RUINED LIVES,** The horror I witnessed inside the women's "liberation" movement' Kate's sister Mallory recounts a meeting of the group that became the National Organization for Women (NOW) that she attended at the invitation of her sister Kate.

[26] McCain, R. Stacy, *Sex Trouble.*

"It was 1969. Kate invited me to join her for a gathering at the home of her friend, Lila Karp. They called the assemblage a "consciousness-raising-group," a typical communist exercise, something practiced in Maoist China. We gathered at a large table as the chairperson opened the meeting with a back-and-forth recitation, like a Litany, a type of prayer done in Catholic Church. But now it was Marxism, the Church of the Left, mimicking religious practice:

"Why are we here today?" she asked.

"To make revolution," they answered.

"What kind of revolution?" she replied.

"The Cultural Revolution," they chanted.

"And how do we make Cultural Revolution?" she demanded.

"By destroying the American family!" they answered.

"How do we destroy the family?" she came back.

"By destroying the American Patriarch," they cried exuberantly.

"And how do we destroy the American Patriarch?" she replied.

"By taking away his power!"

"How do we do that?"

"By destroying monogamy!" they shouted.

"How can we destroy monogamy?"

Their answer left me dumbstruck, breathless, disbelieving my ears. Was I on planet earth? Who were these people?

"By promoting promiscuity, eroticism, prostitution and homosexuality!" they resounded.

They proceeded with a long discussion on how to advance these goals by establishing The National Organization of Women. It was clear they desired nothing less than the utter deconstruction of Western society. The upshot was that the only way to do this was "to invade every American institution. Every one must be permeated with 'The Revolution'": The media, the educational system, universities, high schools, K-12, school boards, etc.; then, the judiciary, the legislatures, the executive branches and even the library system."[27]

This was in 1969, by now, they and their allies have largely succeeded. Their goal of 'promiscuity, eroticism, prostitution and homosexuality' is so successfully accomplished that among the young today, I highly suspect that the sexually normal is outrageously abnormal. Their goals are literally satanic in nature, and their actions can quite comfortably be categorized under what Mueller calls Spiritual Obsession, and likely in his narrow sense of 'those wicked persons whose minds are possessed, filled, and actuated by Satan in spiritual darkness.'[28]

Exhibit B:
Barnard College's Center for Research on Women

"Feminists have very interesting plans for your children. In February 2015, Barnard College's Center for Research on Women announced an "Action on Education" conference that featured a panel:

Dreams of Feminist Education *Tadashi Dozono, Ileana Jiménez, Cheyenne Tobias*

Two teachers of color, both feminist and queer, will share their dreams for feminist education in schools. Moving from theory to action, Ileana and Tadashi work alongside their students using various feminisms such as women of color feminism, global feminism, trans-feminism and queer theory. Their pedagogical practices incorporate restorative and social justice, inspiring

---

[27] Millett, Mallory: **MARXIST FEMINISM'S RUINED LIVES,** The horror I witnessed inside the women's "liberation" movement. https://www.frontpagemag.com/fpm/240037/marxist-feminisms-ruined-lives-mallory-millett

[28] Mueller, John Theodore Th.D., *Christian Dogmatics* 203.

innovative curricula that are intersectional and interdisciplinary. In collaboration with Cheyenne Tobias, feminist artist and Ileana's former student, Tadashi and Ileana will bring us on a visual journey through two different school contexts via the successes they've had and the challenges they face in bringing a feminist vision to their respective classrooms. Calling us to action through their own personal storytelling, Ileana and Tadashi will urge us to consider the role of feminism in schools and the role that schools play in feminism.

What kind of lessons do these self-described queer feminists want to teach your children? Would it be paranoid to suspect that these lessons might involve destroying capitalism and practicing witchcraft? If radical theories are embraced and promoted at elite institutions like Barnard College (where tuition for the 2014-2015 school year was $46,040), shouldn't we expect that there will be an intellectual trickle-down effect, so that these feminist ideas about "social justice" are diffused into the K-12 curriculum? If parents do not "share their dreams for feminist education in schools," what can be done to prevent these queer feminists from teaching their "inspiring innovative curricula" to future generations?"[29]

Essentially, this is every single 'woman's studies' or 'gender studies' department on every campus in every state. These are the crack troops of the NOW plan from Exhibit A.

I could continue, but there is a necessary point that still must be made. If you can go and read critically what is taught in a 'Woman's Studies' class, or go read a feminist blog on tumblr you can find all the rest of this yourselves. I especially recommend R. S. McCain's writing. He is essentially a journalist of insanity. All in all, the 'equal pay for equal work' and the 'equal rights under the law' is a charade of reasonable, worthy goals, propagated by organizations like 'International Woman's Day' to get reasonable

[29] McCain, R. Stacy, *Sex Trouble.*

honest men and women to be feminist supporters in the interests of their own subversive ends.

Many leading feminists (and many Tumblr feminists) are dabblers (or worse) in witchcraft. This has become even more apparent after the election of Donald Trump. Irrespective of your politics, it should be disconcerting to find that Sally Quinn honestly believes, and puts in her memoir,[30] that she successfully hexed people to death. Likewise it should be disconcerting to find out that the Podestas (who are Clinton insiders and very important behind the scenes types) are comfortable with the idea of 'spirit cooking'[31] whether they believe it or not. Or that Tony Podesta collects art that is outright horrifying. Art tells you a good deal about the artist, and an immense amount about the collector. What Tony owns is so bad, that I cannot even bring myself to go get a public domain image and put it in this essay. Suffice it to say that if you steel yourself, a quick (very brief), Internet search will nauseate you. Lastly, we have 'thousands' of witches casting spells on Trump every month.[32]

The upshot is that 'Feminism is Queer'[33] and abortion is an essential sacrament to the feminist leaders, and witchcraft is common. Lena Dunham, you may have heard, said that she regretted never having had an abortion.[34] Students are being taught that heterosexuality is itself oppressive patriarchy. They are being

---

[30] https://www.washingtonpost.com/outlook/sally-quinns-hexes-marital-ultimatums-and-visceral-love-of-her-son/2017/09/08/94694dfe-882b-11e7-961d-2f373b3977ee_story.html?utm_term=.764e5b82e2db

[31] https://www.washingtontimes.com/news/2016/nov/4/wikileaks-john-podesta-invited-to-spirit-dinner-ho/

[32] https://www.vox.com/2017/6/20/15830312/magicresistance-restance-witches-magic-spell-to-bind-donald-trump-mememagic

[33] Feminism is Queer: The intimate connection between queer and feminist theory by Mimi Marinucci A professor at Eastern Washington University. Bio states: **Mimi Marinucci** joined the faculty in 2000. She came to EWU from Temple University in Philadelphia, PA where she completed a Ph.D. in Philosophy and a graduate certificate in Women's Studies. She has a joint appointment, and her time is therefore split equally between Philosophy and Women's & Gender Studies.

[34] https://www.nationalreview.com/blog/corner/lena-dunham-wishes-she-had-abortion/

taught to smash the 'heteronormative gender binary', and the 'imperialist western beauty standards'. Girls are being taught this from before high-school.

If that started to sound a bit like insanity, well, you will also find out, if you peak around the corner into this sector, that many, maybe even most, young feminists deem themselves mentally ill. They have PTSD, or chronic depression, or personality disorders. They brag about their diagnoses.

In the modern LGBT community it is super popular to invent your own pronouns. So that you don't imply a 'gender binary' by speaking about he/she. From the LGBT resource center at University of Wisconsin Milwaukee, we find this helpful guide.

1	2	3	4	5
(f)ae	(f)aer	(f)aer	(f)aers	(f)aerself
e/ey	em	eir	eirs	eirself
he	him	his	his	himself
per	per	pers	pers	perself
she	her	her	hers	herself
they	them	their	theirs	themself
ve	ver	vis	vis	verself
xe	xem	xyr	xyrs	xemself
ze/zie	hir	hir	hirs	hirself

Figure 22: the many and various invented pronouns for students who are LGBTQIAA etc.[35]

The FAQ from the same site has a question: 'Why is it important to respect people's pronouns?' the answer is

> *You can't always know what someone's pronouns are by looking at them. Asking and correctly using someone's pronouns is one of the most basic ways to show your respect for their gender identity.*

[35] http://uwm.edu/lgbtrc/support/gender-pronouns/

When someone is referred to with the wrong pronoun, it can make them feel disrespected, invalidated, dismissed, alienated, or dysphoric (often all of the above.)

It is a privilege to not have to worry about which pronoun someone is going to use for you based on how they perceive your gender. If you have this privilege, yet fail to respect someone else's gender identity, it is not only disrespectful and hurtful, but also oppressive.[36]

As you peer into this lunacy longer an odd item ought to appear. While there are many 'boutique' pronouns, one of the most common is that a single individual will *demand* to be addressed as a plural 'they/them/theirs'. From a Christian perspective we ought to at least entertain the idea that perhaps for some, this is the *correct* pronoun, and that we have a large and growing demon possession problem to match the growing occult fascination.[37] Many in our society are like Rishda Tarkaan; they are calling upon Tash, and many are surprised when Tash comes.

As I look out over the battlefield, I see these feminists, the occult dabblers in in the political world (who may well be more than dabblers), and the current LGBT utter denial of basic scientific biology to be of the same nature. I also think that a time is coming soon that the hard-headed scientists who have often been disparaged in our Christian community as believing in Evolution because they hate God, may well become helpful allies. Not best buddies, since they are not Christian, but at very least allies of convenience. It may also be that it is in this situation that the Word of Truth may be audible to the scientific skeptics. These are often people who hard-headedly desire the truth. Sometimes they desire truth more passionately than those in the Church. The

---

[36] http://uwm.edu/lgbtrc/support/gender-pronouns/

[37] For an account of the problems of demon possession in the Lutheran Church in Madagascar, and a discussion of the issue facing modern America see two books by Robert H. Bennett. 'Afraid: Demon Possession and Spiritual Warfare in America' and 'I am Not Afraid'.

desire for Truth, combined with the fact that they are under direct and unrelenting assault by the LGBT community (professors may get driven out of their jobs by angry mobs when they remark that the human race is a sexually dimorphic species, which means that the males have certain characteristics not shared by the females and vice-versa) may open their ears to the gospel.

This is why I believe that Christians need to reengage the sciences. The sciences are a tool crafted by Christians which will be potentially vital in our upcoming confrontation with the forces described in this essay. These are the people who would willingly show up at a church like mine where women cannot be pastors and homosexuals are called to repent of their sin and beat everyone inside with baseball bats and burn the building to the ground. It is those who follow this ideology who will, sometime as soon as they can get away with it, find a judge to rule that an institution like Concordia University Wisconsin *must* hire the transgender 'woman' who claims to be Christian. The judge will state that it is unconstitutional to discriminate on the basis of race, gender, or sexual orientation, and that since this individual (probably a 'They') claims to be Christian, there is no freedom of religion case to be made. Are we ready for either of these scenarios? Because the cavalry is charging already.

I want to end this book with a verse of one of the best hymns written after 1600: Water, Blood, and Spirit Crying. That and a reminder that we are in a war, a deadly, long campaign, we are members of the church militant, and fight we must.

Water, Blood, and Spirit Crying — LSB 597

Though around us death is seething,
God, His two-edged sword unsheathing
By His Spirit life is breathing
Through the living, active Word.

We live in the church militant. Therefore, we must "be strong in the Lord and in the strength of his might. Put on the whole armor of God, that you may be able to stand against the schemes of the devil. For we do not wrestle against flesh and blood, but against the rulers, against the authorities, against the cosmic powers over this present darkness, against the spiritual forces of evil in the heavenly places. Therefore, take up the whole armor of God, that you may be able to withstand in the evil day, and having done all, to stand firm. Stand therefore, having fastened on the belt of truth, and having put on the breastplate of righteousness, and, as shoes for your feet, having put on the readiness given by the gospel of peace. In all circumstances take up the shield of faith, with which you can extinguish all the flaming darts of the evil one; and take the helmet of salvation, and the sword of the Spirit, which is the word of God, praying at all times in the Spirit, with all prayer and supplication. To that end, keep alert with all perseverance, making supplication for all the saints." (Philippians 6:10-18)

Study Work:

1. Describe the sources of evil in the world.
2. Did you describe evil like a deist describes God? (e.g. is evil an impersonal force for ill, which is almost interchangeable with very bad luck?)
3. Read Koehler, Part IV, XII on the angels (44-47 in my copy)
4. Describe each source of evil in the Christian's life: The Devil, The world, and our own flesh.
5. How is one to treat the devil?
6. What ought to be done with evil?
7. Does demon possession happen today?

Bibliography

Some of these are explicitly cited, other strongly inform my thoughts. All are worth reading, though some must be approached with a critical mind.

Behe, Michel, *Darwin's Black Box: The Biochemical Challenge to Evolution*, Touchstone, New York, 1996.

Bennett, Robert H. *Afraid: Demon Possession and Spiritual Warfare in America*, Concordia Publishing House, St Louis, 2016.

Bennett, Robert H. *I Am Not Afraid: Demon Possession and Spiritual Warfare*, Concordia Publishing House, St. Louis, 2013.

Chesterton, G. K. *The Ballad of the White Horse*, Project Gutenberg, 2013.

Chesterton, G. K. *The Wild Knight and other Poems*, Project Gutenberg, 2018.

Chesterton, G. K. *Eugenics and Other Evils*, Project Gutenberg, 2008.

Day, Vox, *Jordanetics*, Castalia House, Kouvola, Finland, 2018.

Gaiman, Neil, *American Gods*, Harper Collins, 2011.

Hallpike, C.R. *Ship of Fools,* Castalia House, Kouvola, Finland, 2018.

Harold G., Lt. Gen., USA, (ret.) and Galloway, Joseph L. Moore, *We Were Soldiers Once . . . And Young - Ia Drang, The Battle That Changed The War In Vietnam,* Harper Perennial, 1994.

Harstad, Adolph, *Joshua*, Concordia Commentary, Concordia Publishing House, St. Louis.

Kipling Rudyard, *The Gods of the Copybook Headings*

Klavan, Andrew, *The Crisis in the Arts: Why the Left Owns the Culture and How Conservatives can Begin to Take it Back,* David Horowitz Freedom Center, 2014.

Klotz, John W. *Modern Science in the Christian Life*, Concordia Publishing House, St. Louis MO, 1961.

Koehler, Edward W. A., *A Summary of Christian Doctrine*, Concordia Publishing House, St. Louis MO, 1952.

Kuhn, Thomas S., *The Structure of Scientific Revolutions* 4th ed. The University of Chicago Press, 2012.

Lewis, C. S. *Prince Caspian,* Harper Collins, 2002.

Lewis C.S. *The Abolition of Man*, HarperOne, 2015.

Luther, Martin *Commentary on Genesis: Luther on Sin and the Flood* Trans. John Nicholas, Lenker D.D., The Luther Press, Minneapolis MN, 1910.

Lutheran Service Book, Concordia Publishing House, St. Louis Mo.

Meek, Esther Lightcap, *Longing to Know: The Philosophy of Knowledge for Ordinary People*, Brazos Press, Grand Rapids MI, 2003.

Montgomery, John Warwick, *Principalities and Powers,* Canadian Institute for Law, Theology & Public Policy, Inc., 2001.

Morris, Henry M. *The Genesis Record: A Scientific and Devotional Commentary on the Book of Beginnings*, Baker Books, Grand Rapids MI, 1976.

McCain, R. Stacy, *Sex Trouble*, Self-published, 2015.

Mueller, John Theodore Th.D., *Christian Dogmatics*, Concordia Publishing House, St. Louis MO, 1934.

Murray, Scott R. ed. *The Gates of Hell: Confessing Christ in a Hostile World*, Concordia Publishing House, St. Louis MO, 2018.

Parton, Craig, *The Defense Never Rests: A Lawyer's Quest for the Gospel 4th ed.* Concordia Publishing House, St. Louis MO, 2003.

Polkinghorne, John, *Science and Religion in Quest of Truth*, Yale University Press, New Haven, 2011.

Pratchett, Terry, *Hogfather*, HarperCollins Publishers Inc. New York, 1996.

The Holy Bible, English Standard Version® (ESV®) Crossway, 2001.

Tour, James M. Two Experiments in Abiogenesis, *Inference: International Review of Science* 2, No. 3 (2016). http://inference-review.com/article/two-experiments-in-abiogenesis

Tour, James M. An Open Letter to My Colleagues, *Inference: International Review of Science* 3, No. 2 (2017). https://inference-review.com/article/an-open-letter-to-my-colleagues

Veith, Gene Edward, *Modern Fascism: Liquidating the Judeo-Christian Worldview, Concordia Publishing House,* St. Louis MO, *1993.*

Williams, Charles, *The Descent of the Dove,* Regent College Publishing, 2001.

In Christ All Things Hold Together: The Intersection of Science & Christian Theology, CTCR, 2015

CPSIA information can be obtained
at www.ICGtesting.com
Printed in the USA
LVHW031718250121
677445LV00038B/768